AF739553

# Fisherfolk in Cambodia, India and Sri Lanka

This volume studies the coastal and riparian fishing communities of three Asian countries: Cambodia, India and Sri Lanka. It explores issues of migration and movement, gender relations, well-being and nature–society relations common among these communities and studies the impacts of internal and external pressures such as changing state policies, increased market exposure and unstable environmental situations. It also discusses the changes needed to ensure safe migration, social inclusion and the gendered well-being of fishers in these countries, and it identifies the roles that social networks and collective action play in bringing about these improvements.

*Fisherfolk in Cambodia, India and Sri Lanka* presents a rigorously investigated account of the peoples and production systems of some of Asia's most populated and contested but dynamic and productive coasts and floodplains. The book will be of importance to students and researchers of Asian studies, development studies, geography, sociology, migration studies, gender studies and minority studies.

**Ragnhild Lund** is Professor of Geography in the Department of Geography, Norwegian University of Science and Technology. Her research focuses on three broad areas: gender and development (livelihoods, body space, activism, rethinking and reproducing gender and gender-based violence); mobility and migration (mobile livelihoods, development-induced displacement, gendered mobility and Indigenous people); and social geography (cultural encounters, youths, children, friends, house and home and postwar and post-disaster recovery). She has published extensively in international peer-reviewed journals and books and has led several large research projects in Asia.

**Kyoko Kusakabe** is Professor of Gender and Development Studies in the Department of Development and Sustainability, School of Environment, Resources and Development, Asian Institute of Technology (AIT), Thailand. Her main research interests are gender and work, especially labour migration, unpaid work and unpaid care work, the informal economy, fisheries and aquaculture. She has worked extensively with other academics, local NGOs and groups, governments and both international and UN organizations.

**Nitya Rao** is Professor of Gender and Development in the School of International Development, University of East Anglia, Norwich, UK. She has worked extensively as a researcher and advocate in the field of women's rights, employment and education for over three decades. Her research interests include exploring the gendered changes in land and agrarian relations and in migration and livelihoods, especially in contexts of climatic variability and economic precarity. She has conducted research on intra-household dynamics in these contexts, to draw out implications for gendered well-being, with a focus on food and nutrition security. She has published extensively in international peer-reviewed journals and books.

**Nireka Weeratunge** is an anthropologist and research fellow at the International Centre for Ethnic Studies (ICES) in Colombo, Sri Lanka. Her main areas of research are the social and cultural aspects of natural resource use, focusing on gender and livelihood strategies in relation to poverty, vulnerability, well-being and the resilience of rural households in fishing and farming communities. She has a PhD in anthropology from the University of Toronto, Canada.

# Fisherfolk in Cambodia, India and Sri Lanka

## Migration, Gender and Well-Being

Edited by
Ragnhild Lund, Kyoko Kusakabe,
Nitya Rao and Nireka Weeratunge

Routledge
Taylor & Francis Group
LONDON AND NEW YORK

First published 2021
by Routledge
2 Park Square, Milton Park, Abingdon, Oxon OX14 4RN

and by Routledge
52 Vanderbilt Avenue, New York, NY 10017

*Routledge is an imprint of the Taylor & Francis Group, an informa business*

© 2021 selection and editorial matter, Ragnhild Lund, Kyoko Kusakabe, Nitya Rao and Nireka Weeratunge; individual chapters, the contributors

The right of Ragnhild Lund, Kyoko Kusakabe, Nitya Rao and Nireka Weeratunge to be identified as the authors of the editorial material, and of the authors for their individual chapters, has been asserted in accordance with sections 77 and 78 of the Copyright, Designs and Patents Act 1988.

All rights reserved. No part of this book may be reprinted or reproduced or utilised in any form or by any electronic, mechanical, or other means, now known or hereafter invented, including photocopying and recording, or in any information storage or retrieval system, without permission in writing from the publishers.

*Trademark notice*: Product or corporate names may be trademarks or registered trademarks, and are used only for identification and explanation without intent to infringe.

*British Library Cataloguing-in-Publication Data*
A catalogue record for this book is available from the British Library

*Library of Congress Cataloging-in-Publication Data*
Names: Lund, Ragnhild, editor. | Kusakabe, Kyoko, editor. | Rao, Nitya, editor. | Weeratunge, Nireka, editor.
Title: Fisherfolk in Cambodia, India and Sri Lanka: migration, gender and well-being/edited by Ragnhild Lund, Kyoko Kusakabe, Nitya Rao and Nireka Weeratunge.
Description: Abingdon, Oxon; New York, NY: Routledge, 2020. | Includes bibliographical references and index.
Identifiers: LCCN 2020007672 (print) | LCCN 2020007673 (ebook) | ISBN 9780367030476 (hardback) | ISBN 9781003053026 (ebook)
Subjects: LCSH: Fishers – Cambodia – Social conditions – 21st century. | Fishers – India – Social conditions – 21st century. | Fishers – Sri Lanka – Social conditions – 21st century. | Fishers – Cambodia – Economic conditions – 21st century. | Fishers – India – Economic conditions – 21st century. | Fishers – Sri Lanka – Economic conditions – 21st century.
Classification: LCC HD8039.F66 F57 2020 (print) | LCC HD8039.F66 (ebook) | DDC 305.9/6392095 – dc23
LC record available at https://lccn.loc.gov/2020007672
LC ebook record available at https://lccn.loc.gov/2020007673

ISBN: 978-0-367-03047-6 (hbk)
ISBN: 978-1-003-05302-6 (ebk)

Typeset in Sabon
by Apex CoVantage, LLC

# Contents

*List of figures* vii
*List of tables* ix
*List of contributors* xi
*Foreword* xiv
EDWARD H. ALLISON
*Acknowledgements* xvii

1 Fishers on the move: changing livelihoods, gendered entanglements, and well-being 1
RAGNHILD LUND

2 Migration for capital accumulation: changing class dynamics among small-scale fishers on the Coromandel Coast, Tamil Nadu 25
NITYA RAO AND R. MANIMOHAN

3 Adapting to diminishing fish resources in Cambodia: fisheries on the shoulders of women and migrating adult children in fishing communities 41
KYOKO KUSAKABE AND PRAK SEREYVATH

4 Seasonal migration, resource access, contestation, and conflict among fishers on the west and east coasts of Sri Lanka 58
NADINE VANNIASINKAM, MOHAMED FASLAN AND NIREKA WEERATUNGE

5 Female headship and exclusion from small-scale fishing in Eastern Province, Sri Lanka 75
RAGNHILD LUND AND FAZEEHA AZMI

6 To migrate or not: social well-being and gendered household decision-making in fishing communities on the west and east coasts of Sri Lanka 92
NIREKA WEERATUNGE, RAMANI GUNATILAKA, NADINE VANNIASINKAM, MOHAMED FASLAN, DILANTHI KORALAGAMA AND NIRMI VITARANA

7 Fishing in distant waters: issues of identity and well-being among migrant fishers on the west coast of Tamil Nadu 112
NITYA RAO AND C.M. PRATHEEPA

8 Immobility of ethnic Vietnamese fishers and their adaptation strategy in Chhnok Tru village, Chhnok Tru Commune, Cambodia 132
RAKSA SOK AND KYOKO KUSAKABE

9 Mobility in contexts of precarity: kin solidarity and migrant networks among small-scale fishers in coastal Tamil Nadu 147
NITYA RAO, R. MANIMOHAN AND C.M. PRATHEEPA

10 Mobilizing for and against migration: gendered networks, cooperation, and collective action in fishing communities on the west and east coasts of Sri Lanka 164
NIRMI VITARANA, DILANTHI KORALAGAMA, NIREKA WEERATUNGE AND RAMANI GUNATILAKA

11 Gender and power struggle in community fisheries in Cambodia: creating space for women's leadership 181
KYOKO KUSAKABE, PRAK SEREYVATH, LAM DOEURN AND YEM SIVON

12 Livelihoods, migration, and mobility: the distribution of consumption expenditure in fishing communities in Cambodia, India, and Sri Lanka 196
RAMANI GUNATILAKA

13 Small-scale fishers, mobility, and the politics of well-being in rapidly changing Asia 218
RAGNHILD LUND

*Index* 227

# Figures


| | | |
|---|---|---|
| 1.1 | Cambodia (inland and marine fisheries) – Tonle Sap lake and southwestern coast | 17 |
| 1.2 | India (marine fisheries) – Cuddalore District and Kanyakumari District, Tamil Nadu | 18 |
| 1.3 | Sri Lanka (marine fisheries) – Puttalam District, Trincomalee District, Batticaloa District, and Ampara District | 19 |
| 3.1 | Fish catches by small-scale and commercial fishers in Tonle Sap lake and in coastal areas, Cambodia | 43 |
| 6.1 | Marine fish production in Puttalam District and Trincomalee District, Sri Lanka, 1983–2016 | 95 |
| 6.2 | Fishing households in Puttalam District and Trincomalee District, Sri Lanka, 2004–2016 | 95 |
| 6.3 | Economic sector by most important source of household income in study villages, Puttalam District and Trincomalee District, Sri Lanka | 97 |
| 6.4 | Engagement of men and women by employment activity in study villages, Puttalam District and Trincomalee District, Sri Lanka | 98 |
| 6.5 | Migration patterns in study villages, Puttalam District and Trincomalee District, Sri Lanka | 101 |
| 7.1 | Wealth index quintile distribution across households in the Thoothoor village cluster, Kanyakumari District, Tamil Nadu, 2017 | 123 |
| 7.2 | Gendered decision-making in the Thoothoor village cluster, Kanyakumari District, Tamil Nadu, 2017 (mean values) | 126 |
| 7.3 | Men's and women's perception of well-being in 2017 compared with in 2007, Thoothoor village cluster, Kanyakumari District, Tamil Nadu | 127 |
| 9.1 | Comparison of debt sources in Cuddalore District and Kanyakumari District, Tamil Nadu, 2017 | 160 |
| 10.1 | Livelihood support from social networks for women and men in all households in study villages, Puttalam District and Trincomalee District, Sri Lanka, 2017 | 172 |

10.2 Livelihood support from social networks for women and men in fishing households in study villages, Puttalam District and Trincomalee District, Sri Lanka, 2017 175
12.1 Livelihoods of Cambodian households, 2011 201
12.2 Livelihoods of households in fishing communities in Tamil Nadu, 2017 202
12.3 Livelihoods of households in fishing communities on the west and east coasts of Sri Lanka, 2017 203
12.4 Adaptive kernel density estimation of the distribution of real monthly per capita consumption expenditure by earnings sector, Cambodia, 2011 205
12.5 Adaptive kernel density estimation of the distribution of monthly per capita consumption expenditure by migration, fishing households in Tamil Nadu, India, 2017 206
12.6 Adaptive kernel density estimation of the distribution of real monthly per capita consumption expenditure fishing and non-fishing households, Puttalam and Trincomalee districts, Sri Lanka, 2017 207
12.7 Adaptive kernel density estimation of the distribution of real monthly per capita consumption expenditure by type of migration, fishing households in Cambodia 2011 209
12.8 Adaptive kernel density estimation of the distribution of monthly per capita consumption expenditure by migration, fishing households in Tamil Nadu, India, 2017 211
12.9 Adaptive kernel density estimation of the distribution of real monthly per capita consumption expenditure and internal migration for fishing, Puttalam, Sri Lanka, 2017 213
12.10 Adaptive kernel density estimation of the distribution of real monthly per capita consumption expenditure of fishing households in Kanyakumari, Cuddalore, and Puttalam, 2017 214

# Tables


1.1 Overview of the studies in the project Migration and Collectives/Networks as Pathways out of Poverty: Gendered Vulnerabilities and Capabilities amongst Poor Fishing Communities in Asia 15
2.1 Boat types owned by households in the study villages, Cuddalore District, Tamil Nadu, 2017 29
2.2 Occupational profile of villagers in the study villages, Cuddalore District, Tamil Nadu, 2017 30
5.1 Details of the study participants in Batticaloa District and Ampara District, Sri Lanka, February 2018 80
6.1 Migrants, who migrated for fishing and related livelihoods, from fishing households in study villages, Puttalam District and Trincomalee District, Sri Lanka, 2017 100
6.2 Migration of male fishers and accompanying female family members, by ownership of fishing vessel, from study villages, Puttalam District, Sri Lanka, 2017 100
7.1 Education levels of all adult members of households in the study villages, Kanyakumari District, Tamil Nadu, 2017 119
8.1 Profile of the three groups of households in Chhnok Tru village, Chhnok Tru Commune, Cambodia, by response to decreases in fish catches, 2018 137
9.1 Migration destinations for migrants from Cuddalore District and Kanyakumari District, Tamil Nadu, 2017 151
9.2 Reasons for migration from Kanyakumari District and Cuddalore District, Tamil Nadu, 2017 151
9.3 Strength of social networks perceived by women and men in Kanyakumari District and Cuddalore District, Tamil Nadu, 2017 (measured on a 1–5 Likert scale) 159
12.1 Households' livelihoods in Cambodia (2011) and in studied fishing villages in study districts, both in Tamil Nadu, India (2017), and on Sri Lanka's west and east coasts (2017) 199

12.2 Measures of consumption inequality in Cambodia (2011) and in studied fishing households in Tamil Nadu, India (2017), and Sri Lanka (2017) 204
12.3 External and internal migration for work by livelihood strategy in Cambodia, 2011 208
12.4 Labour mobility among studied fishing households in Tamil Nadu, 2017 210
12.5 Household's migration in fishing communities in Sri Lanka, 2017 212

# Contributors

**Fazeeha Azmi** is a senior lecturer at the Department of Geography, Faculty of Arts University of Peradeniya, Sri Lanka. Her research interests are in Sri Lanka and include postwar youths, poverty and livelihood changes, women and migration, internal displacement, postwar development, and tourism. She has published journal articles and book chapters on gender, migration, youth and forced displacement, and tourism.

**Lam Doeurn** is agriculture technical adviser at the Cambodian Institute for Research and Rural Development (CIRD). He has over 10 years of work experience as a field animator, farmer trainer, technical advisor, and project coordinator of different agricultural-based rural development projects and value chain development projects implemented by the Cambodian Centre for Study and Development in Agriculture (CEDAC) and CIRD.

**Mohamed Faslan** is a lecturer at the Department of Political Science and Public Policy, University of Colombo, Sri Lanka. He was previously a researcher at the International Centre for Ethnic Studies (ICES) in Colombo. His main areas of research are ethnoreligious conflict, peace and conflict, fishing disputes between India and Sri Lanka, and Sri Lankan Muslims. He has a master's degree in political science from the University of Colombo.

**Ramani Gunatilaka** is an economist and research associate at the International Centre for Ethnic Studies (ICES) in Colombo, Sri Lanka. She also works as an independent consultant, specializing in the econometric analysis of labour markets, income distribution, education, and subjective well-being, in Sri Lanka, Afghanistan, the Maldives, and China. She has a PhD in applied econometrics from Monash University, Australia.

**Dilanthi Koralagama** is a social scientist and senior lecturer in the Faculty of Agriculture, University of Ruhuna, Matara, Sri Lanka. Her main areas of research are well-being, resource management, and marketing in fishing and farming communities. She has a master's degree in fisheries economics and management from the University of Tromsø, Norway.

**R. Manimohan** is affiliated with the French Institute of Pondicherry, India. He was Research Coordinator with the Fisheries Management Resource Centre (FishMarc), India. His research interests include agrarian relations and the political economy of natural resource management and policy, focusing particularly on land, water and fisheries in Tamil Nadu, India. He has contributed to several papers on these themes published in national and international journals. He has also taught economics and research methodology at the postgraduate level.

**C.M. Pratheepa** is a researcher and professional social worker. She has designed and implemented various welfare programmes for community well-being, including building the capacities of a network of NGOs in Tamil Nadu to work towards achieving sustainable development goals. She is also a research associate at the Fisheries Management Resource Centre (FishMarc), India. She has published several papers in national and international journals.

**Prak Sereyvath** is a guest lecturer at Prek Leap National College of Agriculture, Cambodia. He was cofounder of the Cambodian Centre for Study and Development in Agriculture (CEDAC). He founded CIRD (Cambodian Institute for Research and Rural Development) in 2009 and currently works as the director of CIRD. Apart from organizational management tasks, he has also designed, managed, and provided support to several agricultural development projects and research projects funded by different national and international development partners and governments.

**Yem Sivon** is a lecturer at the Prek Leap National College of Agriculture, Cambodia, and vice chief of the college laboratory. He also works as part-time consultant at CIRD, conducting field studies and providing training to farmers and leaders of farmer organizations on both agricultural techniques and socioeconomic subjects.

**Raksa Sok** is a project leader at Creative Green Design, Phnom Penh, Cambodia, where she is responsible for social research and environmental impact assessments. Earlier, she was affiliated to the Cambodia Development Research Institute (CDRI) through work on a project related to water governance.

**Nadine Vanniasinkam** is a researcher at the International Centre for Ethnic Studies (ICES), Colombo, Sri Lanka. Her research interests include interreligious relations and conflict, nationalism, and minority identity politics, with gender as a cross-cutting focus. She has a master's degree in applied linguistics from the University of Melbourne, Australia.

**Nirmi Vitarana** is a development practitioner and is working as a programme manager in the Research, Learning and Program Strategy of the Asia Foundation's Subnational Governance Program in Sri Lanka.

She was previously a researcher at the International Centre for Ethnic Studies (ICES), Colombo, Sri Lanka. Her research interests include social networks, the well-being of people with disabilities, and the role of governance in promulgating social justice. She has a master's degree in sustainable international development from the Heller School for Social Policy and Management, Brandeis University, USA.

# Foreword

This landmark book examines the intersection of some of the most important issues confronting contemporary society. It does so through a focus on the coastal and riparian fishing people of three Asian countries – Cambodia, India and Bangladesh – but the questions it asks and the issues it engages apply beyond that already-large, important, and complex arena. The first of the book's themes is the migration and mobility of large numbers of people. Political turmoil, economic crisis and opportunity, and environmental change are driving more people to move, seasonally, temporarily, or permanently, with traditional patterns of fisherfolk's mobility altered by the search for livelihoods beyond the sector. One of the book's central questions is whether migration offers a route out of poverty for fishing communities. The anxieties and hopes of those who choose (or are compelled) to move are narrated in several of the chapters here. These stories have universal significance at a time when the poor are met with fear and hostility at the borders of wealthier nations, even as their labour is needed to support the lifestyles of those nations.

Second, this book engages deeply with gender at a time when progress towards equal rights for women has faltered in some of the world's major nations. The gendered experiences of mobility and change and the relationships among people seeking a decent and dignified life in the midst of much upheaval are critical questions in each chapter here. Moreover, the approach taken by the authors reflects new intersectional ways of unpacking experiences that are shaped by gender, class, and race.

Third, several of the chapters here explore the human quest for well-being. They present universally resonant stories of people's search for improved economic stability, material comfort, good social relations and standing, and a sense of dignity, contentment, and joy. This search for well-being takes place amid increases in both vulnerability and opportunity. Sources of vulnerability in the study area include natural disasters such as the Asian tsunami of 2004 and civil conflict. Both of these greatly affected the fishing communities of Sri Lanka, for example. As inequalities deepen and growth-led economic development fails to curb dangerous environmental change, there is renewed interest in understanding how well-being is achieved and how relational and non-material components can be supported.

Lastly, this is a meditation on society–nature relationships, an ethnographically lit window on the Anthropocene at a time of existential concern about climate change and the intensifying use and abuse of natural resource systems. While it is too simple to say that people migrate when resources become scarce, this has certainly been a factor in driving labour migration in Cambodia, the focus of some of the chapters here. Several chapters explore the role of institutions at multiple levels in enabling or constraining migration decisions and other measures to adapt to environmental change. The concluding chapter points to the need for strengthening community-led institutions for natural resource management.

The breadth and complexity of the issues covered in this volume could easily overwhelm both researcher and reader. Instead, the authors have produced a rigorously investigated and highly readable account of the peoples and production systems of some of Asia's most populated and contested but dynamic and productive coasts and floodplains. The major themes are woven together in a series of case studies that collectively build a rich picture of the lives of millions of people who are adapting to change. This coherence is in part due to the authors' rigorous adoption of a unifying interdisciplinary research framework, drawing principally on frames of analysis from human geography, gender studies, development studies, and social anthropology. Of particular note, conceptually and methodologically, is that this book moves forward the analysis of gender in the fisheries sector. For much of the past 20 years, gender studies in fisheries have been preoccupied with counting women and identifying their roles in the 'ecology' of the fishery system. This has been tremendously valuable in raising the visibility of women in fisheries research and policy, but it has not always told us much about the people behind the labour statistics and income inequalities. This book enables its readers to better understand fisherfolk's motivations, fears, aspirations, and choices. This is surely important in attempts by governments, NGOs, and donors to work with them to improve resource stewardship and well-being.

The 13 chapters in the book explore the entwined issues of migration and movement, gender relations and gendered outcomes, well-being, and nature–society relations in three Asian countries: Sri Lanka, India, and Cambodia. There are several reasons why a focus on coastal and riparian communities engaged in small-scale fishing provides the ideal arena to address the book's themes. Coastal fisherfolk are often mobile or migrant, sometimes to follow fish resources and other times because they have no stable right to land. The work they do is usually (but not always) highly gendered, and there are clear class divisions in communities between those who own assets such as boats, motorized transportation and processing plants or who have the capital to buy large volumes of fish to trade and those who work as labourers on boats, at fish landing areas, or in processing and trading networks. Fishers are also sometimes ethnic minorities or low caste groups or undocumented migrants and often marginalized in land-based societies. Although fishers often report a strong sense of identity, community solidarity, and satisfaction with their work, they also experience social exclusion and marginalization from land-based rural or urban societies. Earlier, the job satisfaction of women working in fisheries was largely unknown, while job satisfaction of men working in fisheries has been much studied by maritime anthropologists and promoted as a major driver of fishers' choice to stay in declining and unprofitable fisheries. As people directly dependent on a functioning and flourishing environment, fisherfolk are also important in both warning us of the costs of destructive nature–society relationships and helping us see how more-sustainable relations might be fostered.

The authors of this book are mostly from the countries where the research was conducted and most also work in research organizations in Asia. Their knowledge of the histories, cultures, and societies of the places they write about is extensive, and the relevant aspects of them are well summarized here. Their readership is likely to include their peers, both regionally and internationally, in the social sciences and humanities and in interdisciplinary areas such as development studies. It has been my pleasure to have worked with some of the authors and a privilege to get to know the work of the others through reading this excellent book. I hope my fellow fisheries researchers, managers, and policy advisers will also read it. We have been guilty of focusing our scholarly attentions more on the fish than on the people who catch, process, and trade them. Consequently, we know more about the migration patterns and lifecycle of sardines and tuna than we do about the movements and lifeworlds of the people who catch them. As governments increasingly talk of devolving responsibilities for fisheries governance back to the communities that traditionally self-regulated their fishing activities and as development actors seek to support people to adapt to a changing world, sustaining coastal resources and livelihoods will require an understanding of and engagement with those whose voices are heard in this fine book.

Edward H. Allison
School of Marine and Environmental Affairs
University of Washington
Seattle, US

# Acknowledgements

Two people facilitated the research on which this book is based. Our sincere appreciation and respect go to the following people. Postdoctoral researcher Fazeeha Azmi facilitated the group's work by setting up a Facebook page, posting articles, and reviewing policy documents, in addition to drafting a book chapter and article on female household heads in India and Sri Lanka. She successfully presented her work at conferences in France and New Orleans, Louisiana. Dr Thomas Sætre Jakobsen facilitated our book application and contributed highly valuable academic insights to the GAF7 conference, seminars, and workshops for the wider project team and specifically to the Trondheim team. We also thank Catriona Turner for copyediting the book manuscript in an excellent manner. Additionally, we thank Ms Karunanayake Nimali Priyangani Piyatissa for preparing Figures 1.1–1.3.

Cambodia:

We thank the members of the Fisheries Action Coalition Team (FACT) for their support in providing information on community fisheries (CFs) in Cambodia and linking the study team to the CFs. The leaders of the CFs are acknowledged for facilitating our study team's meeting and interviews with CF members and for actively engaging in discussion with the members of the study team. The interviews were also facilitated by village heads and commune chiefs in the study areas. We sincerely thank all of the CF member participants who spent time sharing their ideas during the interviews. Mr Ly Vuthy, the director of the Community Fisheries Department at the Department of Fisheries Administration is acknowledged for his inputs during the workshop. Mr Zangta Sang helped Raksa Sok with data collection. Mr Thak, the head of the Vietnamese association in Chhnok Tru Commune, facilitated interviews with ethnic Vietnamese participants. We thank Veena N. for her comments on the chapters and for her help, along with Parichart Khammeerak, in the administrative management of the project at the Asian Institute of Technology. We also thank staff members at the Cambodian Institute for Research and Rural Development (CIRD) for logistical support for the project. Our appreciation also goes to Carlos Sobrado for making available the data on consumption aggregates derived from the Cambodia

Socio-Economic Survey 2011, which were used for the World Bank's Cambodia Poverty Assessment 2013.

India:

Several people and institutions have supported the research in India. First, we thank the people in the community for sharing their stories with us and thank the field investigators (C. Chinnaraj, S. Sridevi, Z. Thabiba, P. Vanaja, S. Vivek, S. Vishnu, and Anand) for their dedicated work. At the village level, several people assisted us in gaining access to the communities. In Kanyakumari District, we particularly wish to thank the pastors in the three study localities – Fr Bebinson, Fr Selvam, and Fr Gildus – as well as *anbiam* organizers Freeda, Sherafi, Maria, and Helen. In Cuddalore District, we were assisted by the headmaster of the Government School in MGR Nagar, the secretary of the fisher cooperatives in the other villages, and Elangovan and Velvizhi of the M.S. Swaminathan Research Foundation. Without the insights and guidance of Vivekanandan and Ephrem of FishMarc, the partner organization in India, and administrative support from Prakash, the research would not have been possible, so we are grateful to them. Finally, we wish to thank Ahana Lakshmi, Clare Shelton, and Narayani Subramaniam for some assistance in reviewing the literature. Clare Shelton is also thanked for preparing Figures 7.1–7.3.

Sri Lanka:

We express our deep appreciation to the representatives of fisheries societies, community and religious leaders, and the people in the two study villages in Puttalam District and the two study villages in Trincomalee District, for their support, engagement, and participation in the Sri Lankan study of coast-to-coast seasonal migration. We wish to especially thank the award-winning young filmmaker Mathavan Maheswaran for training four teams of youths in the study villages as part of a community pilot project and the young filmmakers for their efforts in producing four participatory video documentaries on their livelihoods, migration, and future aspirations. Our thanks are also due to the fisheries officers at national level and in Puttalam District and Trincomalee District and the administrative officers from both Mundel Divisional Secretariat and Arachchikattuwa Divisional Secretariat in Puttalam District and Kuchchaveli Divisional Secretariat in Trincomalee District for sharing their knowledge, providing statistical data, and participating in stakeholder workshops. We thank the Kandy Consulting Group (KCG) and its senior manager of research, Dammika Herath, for conducting the quantitative household survey and ensuring the timely delivery of a good database. The study was made possible due to the abiding interest and support of the executive director of ICES, Mario Gomez, to whom we owe our gratitude. Additionally, we thank Sakeena Alikhan for the preparation of Figure 6.5.

For the study of female household heads, we thank late Dr M. Varnakulasingham for his support in finding fieldwork assistants and for organizing a seminar at the Department of Geography, Eastern University, Sri Lanka.

We express our deep appreciation of Ms Jancy Mauri for interview assistance, transcripts, and assistance with logistics in the field. She carried out the work with the utmost dedication and facilitated access to the villagers in Batticaloa District and Ampara District. Last but not least, we are grateful to the villagers we met during the research and thank them for sharing their life stories with us.

# 1 Fishers on the move

## Changing livelihoods, gendered entanglements, and well-being

*Ragnhild Lund*

### Introduction

This book presents findings from the cross-country research project Migration and Collectives/Networks as Pathways out of Poverty: Gendered Vulnerabilities and Capabilities amongst Poor Fishing Communities in Asia. The project was led by Ragnhild Lund (Norway), with principle investigators Kyoko Kusakabe (Cambodia), Nitya Rao (India and the UK), and Nireka Weeratunge (Sri Lanka) and their respective research teams and partner organizations: the Cambodian Institute for Research and Rural Development (CIRD), the Fisheries Management Resource Centre (FishMarc), and International Centre for Ethnic Studies (ICES). The research was conducted between March 2016 and July 2019 and studied the movements and well-being of fishers in South and Southeast Asia and how their movements were being affected by changing state policies, increased market exposure, and unstable environmental situations.

In recognizing social identity and representation, our project started from the premise that the more than 100 million people who rely on small-scale fishing for their livelihood – of whom approximately 90% are based in Asia – are too important to be left unrecognized (FAO, 2019a). The small-scale fisheries sector is typically firmly rooted in local communities, traditions, and values. Many small-scale fishers are self-employed and provide fish for direct consumption within their households and/or communities. Women are significant participants in the sector, particularly in relation to post-harvest and processing activities (FAO, 2019b).

The project was intended to provide a critical understanding of extensive changes in livelihood strategies that have taken place in fishing communities in recent decades in the context of depleted aquatic resources, technological change, climate change, and social conflicts and how these processes have impacted poverty and well-being. Migration and mobility in fishing communities in Cambodia, India, and Sri Lanka have been motivated by stresses such as seasonal or unpredictable weather patterns, a gradual decline in fish resources, the need to earn a living in the off-season for fishing, and shocks, including fuel price hikes and natural and human-caused disasters, such as the tsunami of 2004 and violent ethnic conflict. Despite such stresses and shocks, most fishers and their household members stay in fisheries-related livelihoods, partly because they are relatively more lucrative than agricultural livelihoods and partly because of the lack of other viable options. Questions remain about changes in gender roles and relations, the resilience of fishing as a way of life and an identity for older and younger generations, and the potential social polarization among fishers who have different technological assets.

Fishers have always been mobile, and seasonal fishing still takes place in the study countries. According to Njock and Westlund (2010), migration is a strategy used by fishing communities to secure livelihoods. Migrant fishers move both internally (domestically) and externally (internationally). Because of the depletion of fish resources and coastal environments, large-scale fishing, the after-effects of the 2004 tsunami, the presence of conflict in the study countries, and transboundary mobility among fishers are increasing. However, the migration of fishers today is not simply motivated by the seasonal availability of fish but also triggered by diverse factors (Islam and Herbeck, 2013; Kramer et al., 2002), including environmental factors such as droughts and floods (Allison and Ellis, 2001; Badjeck et al., 2010; Heinonen, 2006); opportunities created by environmental, economic, and political circumstances (Islam and Herbeck, 2013; Overå, 2005); the rapid expansion of aquaculture and structural economic change (Betcherman and Marschke, 2016); changes in the global economy (Njock and Westlund, 2010); and changes in fish production chains and markets (Gopal et al., 2014; Kusakabe, 2014). With respect to all of the aforementioned factors, migration is gendered and embedded in socioeconomic, cultural, political, and environmental contexts (Hapke, 2001; Weeratunge et al., 2010).

Thus, it follows that while migration may be aspirational, it may also be forced due to distress (Kadfak et al., 2012), and migrants as 'gendered subjects' face different forms of inclusion and exclusion in contexts of scarcity and conflict (Hapke, 2017; Hapke and Ayyankeril, 2018b). When male fishers migrate seasonally, they do so either alone or accompanied by their wives, which has implications for both the migrant male fisher and his wife, because they have to negotiate different social structures to fulfil their well-being expectations in both the places to which they migrate and the places that they have left. As women contribute mainly to post-harvest fisheries activities, the migration of male fishers has implications for women's fisheries-based livelihoods. Aswathy and Kalpana (2018) show how men's migration to the Gulf countries from a fishing community in India forces women with fisheries-based livelihoods to abandon their vocation. Furthermore, new actors are getting involved in the fish trade and production. Thus, men and women are not homogeneous groups but internally differentiated according to their socioeconomic position and marital status. In this book, we study the gendered well-being impacts of fishing households in light of these complex and dynamic contexts. Gender roles must be understood in the context of enabling and constraining structures and external factors, as well as individual goals, values, and aspirations (Britton, 2012; Hapke and Ayyankeril, 2018b).

Central to our work is a comparative understanding of the extent of vulnerability and well-being among fishers in three countries: Cambodia, India, and Sri Lanka. Although migration and/or mobility between different coasts and fishing sites offsets some of this vulnerability, it also brings vulnerabilities to migrant fishers and their households who live in temporary fishing camps without basic amenities and who face conflicts with host communities or who become caught up in transboundary disputes. Moreover, household members who work as migrant labourers overseas face a different set of vulnerabilities that are connected with international labour migration, such as malpractices in recruitment and contract negotiation, poor working and living conditions, and issues of reintegration. We examine social relations at different institutional levels, including a high level of privatized and market-driven institutions in India, relatively more community-level institutions in Cambodia, and the smaller-scale and individual basis in Sri Lanka. In terms of gender relations, there is variation in how women's roles have changed in fishing, ranging from immobility to increasing mobility, given that some women have been tied to domestic work in the home, while others have had to look for alternative work to fishing.

In this book, we look at the different gains and costs of migration affecting women and men and those affecting ethnic and caste groups in sending and receiving communities, and we look into what changes are needed to ensure safe migration, social inclusion, and the well-being of fishers in the three study countries. First, we aim at understanding the gendered well-being and identity of small-scale fishers and specifically ask which approaches are the most effective in enabling the most vulnerable fishers to improve their well-being. Second, we ask how migration may help both women and men in fishing households to exit poverty and improve their well-being. Our understanding of poverty ranges from a focus on economic poverty to one on multiple deprivations, to unravel contextual factors that produce vulnerability and precarity at varying scales, including among male and female fishers, the local fishing communities, and fishers' links to the outside world. Third, we identify the roles that social networks and collective action play in facilitating migration, reducing vulnerability and improving fishers' well-being. We hold that collective action both enables fisherfolk to develop secure livelihoods and provides insights into their attempts to secure livelihoods and maintain their identities as fisherfolk. The individual chapters address the three issues in different ways.

This introductory chapter henceforth begins with reflections on the significance of small-scale fishing at various scales before it goes on to describe the vulnerability of fishers, the modernization and mechanization of fishing, and resource depletion, due variously to decreasing fish catches, climate change, and natural disasters. These are the multiple contextual forces that bring about vulnerability, even precarity, to fishers and their communities. The chapter continues with a presentation of how the individual authors have engaged with theory and the key concepts of the study, and thereafter, it presents the methodology and methods used. The final section outlines the structure of the book.

## The context

At the macro-level, globalization suggests rapid change and connectedness at a number of important levels that condition the lives and work of small-scale fishers. Local economic, political, and cultural contexts respond quickly to processes arising elsewhere, with almost instantaneous market adjustments that affect lives. With regard to the situation of fisherfolk in the Global South, scholarly attention has only recently begun to focus on their shifting livelihoods and well-being as a result of global environmental and sociopolitical changes that have led to increasing incidences of disasters, with implications for mechanization, access to flexible capital, and wider agrarian systems.

A number of scholars have identified small-scale fishers as constituting a vulnerable and sometimes socially excluded group (Bavinck and Johnson, 2008; Béné, 2003; Béné et al., 2016; Johnson et al., 2018; Salagrama, 2006; Weeratunge et al., 2014). However, almost all of these scholars have emphasized that small-scale fishers are not among the poorest of the poor, and they challenge the view of widespread poverty among fishers. However, fishers' increasing vulnerability and potential ill-being (precarity) show that previous measures and policies have failed to address the problems they face. Small-scale fisheries have only recently received more recognition from the Food and Agricultural Organization (FAO) of the United Nations (UN) and from researchers and other stakeholders, most prominently illustrated by the international research network Too Big to Ignore (Johnson et al., 2018). Furthermore, discussion of the UN's Sustainable Development Goals (United Nations, 2019), the FAO's *Voluntary Guidelines for Securing Sustainable Small-Scale Fisheries in the Context of Food Security and Poverty Eradication* (Food and Agricultural Organization of the United Nations, 2015), and 'blue economy' initiatives has set an agenda for future policy formulation for the fisheries sector. There is much to worry about when it comes to the future of small-scale fisheries, given the increasing competition over fishing grounds and shifting state objectives towards urban consumers, with attendant policies geared towards enabling economies of scale in fishing, even at the risk of resource depletion.

Three contingent and partly overlapping contextual forces are making traditional fishing communities vulnerable today:

1 Changing land use patterns
2 Technological innovation and adaptation
3 Environmental degradation, climate change, and resource depletion.

While all of these forces pertain to phenomena occurring both inside and outside fishing communities, the force pertains to how land resources previously used for household social reproduction or for subsistence fisheries are becoming expropriated for other purposes, such as tourism, aquaculture, industries, military bases, real estate, and cash crops (on land grabbing in Southeast Asia; see Hall et al., 2011). Such changes exacerbate the impacts of calamities that already affect fishing communities, such as depleting fish resources and climate change. This raises serious concerns about the future role of fishing as a way of life (Trimble and Johnson, 2013). In addition, new political tensions have increased due to overfishing (e.g. Indians trawling in Sri Lankan waters). In other cases, geopolitical changes have impacted access to water resources (e.g. access to Tonle Sap lake has decreased due to dam construction in China). Furthermore, hotel development, industrialization along coasts, and other modernization projects are putting pressure on coastal communities and small-scale fisheries, thus increasing their vulnerability.

Modernization (relating to new policies, the redistribution of access to fish resources, improved vessels, and infrastructure) and mechanization (pertaining to motorization and improved fishing technologies) are core themes in the literature on small-scale fisheries. These changes, referred to as the Blue Revolution, have created distinct divisions and inequalities among fishers, which in turn have affected relations of production in India (Hapke, 2001; Salagrama, 2006; Vivekanandan and Kasim, 2011), in Sri Lanka (Wijayaratne and Maldeniya, 2003), and in Cambodia (Kusakabe, 2017; Sneddon, 2007). Sneddon (2007) discusses how neoliberal policies – in particular, processes of privatization and capital accumulation – have increased socio-ecological tension in Cambodian freshwater fisheries. In India, research has uncovered how modernization and mechanization in fisheries have altered patterns of class, caste, and gender (Hapke, 2001; Ram, 1991).

Studies on the impacts of mechanization and motorization on fisheries in South India have revealed the negative effects of these processes on small-scale fishers (Johnson et al., 2018; Vivekanandan and Kasim, 2011). The availability of and access to assets, particularly boats and nets, impact the opportunities available to poor fishing households. The thrust for modernization in fisheries is continuously supported by both national and state fisheries policies, including the latest national fishery policy in India, which emphasizes reforms in commercialization and export practices (Department of Animal Husbandry Dairying and Fisheries, 2017). With regard to Sri Lanka, Amarasinghe (1989) discusses the impacts of technological change, focusing on risk and patronage in the fisheries sector, and demonstrates how mechanization has resulted in the overexploitation of fisheries resources in South Asia, in turn eroding the quality of life and livelihoods of small-scale fishers in the region.

Adding to the changes brought about by modernization and mechanization, recent environmental changes – induced by climate change and natural hazards such as tsunamis and floods – have increased the vulnerabilities of many fishers in the study countries by reducing their access to fisheries resources. In Cambodia, climate change has significantly increased economic and social vulnerability (Allison et al., 2009), reducing fishers' capacity to adapt and weakening their overall resilience (Nuorteva et al., 2010). The Indian fisheries sector has similarly suffered the impacts of environmental change, which have resulted in declining fish stocks, fish catches, and livelihoods, thus affecting the national economy.[1] In Sri Lanka, the impact of environmental change on marine habitats and coastal communities has reduced small-scale fishers' access to coastal areas, disturbed the equilibrium of their social order, and pushed them to seek alternative livelihoods (Athulathmudali et al., 2011). In addition to these environmental changes, the 2004 tsunami in India (Jeeva et al., 2011) and Sri Lanka (De Silva and Yamao, 2007) had a devastating socioeconomic impact on coastal fishing communities.

## Theory and concepts

The theoretical approach to our research has been and is eclectic. For the research presented in this book, we used mobility and migration perspectives to unravel temporal and spatial movements; gender studies to reveal how men and women are differentially impacted by recent changes in fishing communities; the social well-being approach to unpack material, relational, and subjective well-being; and political ecology to unpack nature–society interconnections and power relations at multiple scales. The combination of these theories enabled us to look at the multiple factors that might contribute to well-being among migrants and fishing groups.

### *Mobility and migration*

Mobility and migration are intended to lead to the fulfilment of people's aspirations and strategies to overcome vulnerability. Migration is a complex phenomenon, with multiple drivers, pathways, and outcomes. Crona and Rosendo (2011) claim that although migration is an important livelihood strategy practised in fishing communities, it has not received adequate attention in marine governance and coastal development policies. Recent research has moved beyond the economic drivers of migration in order to tease out specific contextual factors – a combination of ecological, socioeconomic, and institutional drivers that motivate fishers to move. These divers include near-shore resource depletion, changes in fishing technologies and practices, the effects of migrant networks, and issues of social identity (Kraan, 2009; Krishnan, 2010; Marquette et al., 2002; Overå, 2001).

Migration is an integral part of the fishing profession and is shaped by the spatial and temporal movements of marine stocks; hence, fisheries-related migration can involve temporary or permanent movements of fishers to other coastal destinations linked to their fishing activities (Abobi and Alhassan, 2015). Overå (2001) argues that migration facilitates resource usage in new ecological niches, enabling small-scale fishers to increase their harvesting of fish in response to seasonal and long-term 'booms' in the ecological system, connected with upwelling in the ocean and the mobility of fish species. The role of translocal social and institutional networks in initiating and sustaining fishers' migration to expand social and livelihood spaces is emphasized too (Kraan, 2009; Overå, 2001). Migration, as understood in this book, feeds into ideas of mobility as both temporal, involving intragenerational or intergenerational change in both social and economic standing, and spatial, ranging from daily trips to longer voyages. In this book, our approach follows a similar strand of ideas.

### *Gender perspective*

A gender perspective runs throughout this book. We show that processes of change are not gender-neutral. The Indian fisheries sector alone, including coastal, inland, and aquaculture fisheries, employs more than 14 million people directly and many more indirectly, and women constitute more than 66% of the total workforce in post-harvest fisheries (Department of Animal Husbandry Dairying and Fisheries, 2017). Despite this, their aspirations of well-being are continuously challenged due to changes taking place both inside and outside the sector. Although women make important contributions to fisheries and household survival, their roles are undervalued and marginalized (Britton, 2012; Hapke, 2001; Hapke and Ayyankeril, 2018a; Kusakabe, 2014). As inequalities among men and women intersect through class, caste, and gender relations, gender roles are constantly renegotiated and reworked in the context of modernization, economic transformation, ecological crisis, local political power structures, and the social constructions of femininity and masculinity (Hapke 2001; Ram, 1991). Hapke (2017), with reference to the situation in Kerala, India, argues that recent social, economic, political, and legal changes have worked to the detriment of women, lowering their status. In another article, Hapke and Ayyankeril (2018a) document how globalization plays out differently for men than for women in fishing communities as a result of different configurations of gender, work, culture, identity, and economy and in how households and communities are connected to fish economies at different scales. In our research, a gender lens has helped us to understand how fishing communities that often constitute tightly knit kin-based social networks work under changing environments and resource depletion, how the gender-based division of labour is changing, and how fishers adapt their strategies in response to both internal and external pressures and changes.

### *Vulnerability and precarity*

Vulnerability and precarity refer to the dynamics between people living and subsisting partly through fishing, the changing political objectives of the state, and the effects of increased market exposure (Ferguson, 2015; Hall et al., 2011; Murray Li, 2011, 2014, 2017). To understand these dynamics, Rigg et al. (2016) point to how increased exposure to markets may have contradictory outcomes for the rural poor, because states and development agencies often perceive that the reason for rural poverty is based on the lack of market access and relative isolation of agrarian livelihood earners. From this perspective, investments in roads, education, and technology and increased access to credit are important for overcoming these inherited vulnerabilities (e.g. lack of credit, insufficient fishing grounds, spatial isolation, and marginalization associated with caste and gender). However, although such efforts might reduce 'old vulnerabilities' (e.g. higher exposure to the vagaries of weather due to a lack of technology, dependence on monopsonistic traders), they might simultaneously produce 'new precarity' (e.g. expulsion from fishing grounds, unsustainable levels of debt, market dependency, increased inequality, increased out-migration, left behind children and elderly, and increased competition) (Rigg et al., 2016). We draw on this insightful contribution concerning the effect of market exposure and shifting state objectives on the vulnerability–'precarity' dynamic and add a gender-sensitive and context-specific understanding of well-being in order to unravel how agency and contexts of conditioning shape the possibilities for small-scale fishers to live well amid economic, technological, and spatio-political restructuring.

### *Well-being*

Well-being is broadly understood as a measure of human development (Sen, 1999) that draws attention away from a money-metric measure of poverty and towards a multidimensional view. This discourse includes various dimensions, such as emotions, relationships, and material outcomes in health and education. Several chapters look at these broader aspects of well-being rather than income alone. Like Johnson et al. (2018), in this book, we use well-being as an analytical lens to understand the outcome of successful negotiation and exit from the multidimensional elements of poverty and vulnerability (Mosse, 2010). This understanding of well-being is juxtaposed with small-scale fishers' own conceptions and pursuits of what they deem a good life, to understand how well-being is undermined or strengthened by existing practices and contexts of conditioning (Johnson et al., 2018).

Some of the contributing authors in this book use the social well-being approach (White, 2017)[2] in particular, to analyse material (economic and ecological), relational (social and political), and subjective dimensions of livelihood strategies and outcomes, focusing on the ability of fisherfolk to live well according to their own values (Bavinck and Vivekenandan, 2010; Britton and Coulthard, 2013; Coulthard, 2015; Coulthard et al., 2011; Coulthard et al., 2014; D'Anna and Murray, 2015; Weeratunge et al., 2014). In this three-dimensional elaboration of social well-being, *material* concerns encompass practical welfare and standards of living (e.g. income, wealth, assets, environmental quality, physical health, and livelihood concerns). *Relational* aspects include relations of love and care, networks of support and obligation, and social, political, and cultural identities, including relations with the state and formal structures, which shape the scope for personal action and community influence. *Subjective* aspects span notions of self, individual hopes, fears and aspirations, expressed levels of satisfaction or dissatisfaction, trust, and confidence, among others (White, 2008, p. 11). This perspective enables us to understand the motivations underlying the pursuit of well-being as a process, especially to comprehend its non-material dimensions (Weeratunge et al., 2014).

### *Collective action and networks of support*

Collective action and networks of support and obligation seek to transform the terms of engagement between differently positioned people and groups, pertaining to gender relations; divisions of labour; relations with peers, community institutions, and collectives; and relations with state agencies and market mechanisms. Our empirical material focuses on awareness, identity, relationships, and responsibilities in order to theorize about engagements and collective networks. Specifically, it is used to explore gendered processes whereby social and cultural norms and expectations mediate how male fishers and female fishers understand and give meaning to these factors. Networks and collectives of fisherfolk mediate social exclusion and inclusion across institutions. By focusing on their roles and responsibilities, we answer questions about the effects of policies and assess whether the policies adequately challenge the status quo, particularly in their articulation on the ground.

In analysing social networks, we draw on the work of Bourdieu (1977), who looked at social capitals as enabling actors to access differential benefits in relation to their inclusion in such networks. Other perspectives connected to social relations, such as the livelihoods approach and social exclusion, help to unpack agency, inequality, and gender differences more explicitly. Some contributors to this book engage with political ecology (Blaikie, 1999), which looks at the relationship between political, economic, and social factors and the environment, focusing especially on changes, such as land degradation or resource depletion, linked to unequal access to resources by different social groups. The political ecology perspective helps to understand the institutional inequalities in access to resources.

The contributors to this book approach the various theories and key concepts differently, and by doing so, they document multiple meanings and contextual variations from one place and country to the other. In the following section, I show how they are intertwined and used in the analyses in the different chapters.

## Individual chapter contributions

### *Mobility and migration*

In the individual chapters, examples of mobility are those of fish vendors and fish auctioneers who commute daily to sell or auction fish (Chapters 5 and 7) and those who engage in temporary labour on fishing grounds (Chapter 2) or in factories in cities (Chapter 3). In other cases, new fishing patterns lead to the immobility or domestication of female fishers (Chapters 7 and 8). Migration is either seasonal (Chapter 6) or for longer stays overseas for fishing (Chapters 2 and 4) or other types of work (Chapters 3 and 5).

It is commonly assumed that fishing is a way of life and that fishers who stay away from fishing for a period of time ultimately return to the practice. However, researchers have argued that small-scale fishers are beginning to regard fishing as undesirable for their children and are encouraging them to get jobs in industry or to migrate to cities or abroad (Trimble and Johnson, 2013). In this book, Nitya Rao and R. Manimohan (Chapter 2) explore the gendered vulnerabilities and well-being impacts of migration in the context of wider transformations in fisheries and coastal landscapes in India. Increasing competition for limited, open access resources has pushed fishers towards the greater capitalization of fisheries, with those able to invest reaping the benefits and others being reduced to a class of labourers. A number of young men from fishing communities migrate to raise capital for investment in boats and allied technologies. New forms of organization are emerging to facilitate the collective ownership of boats, enabling participation in appropriate consumption activities. Dowry has emerged as both a source of capital and a source of status.

Kyoko Kusakabe and Prak Sereyvath (Chapter 3) address how migration is one way out of poverty in Cambodia. When fish resources decrease, villagers start to diversify their livelihood options into agriculture, or they engage in non-farming and non-fishing activities, borrow money, or migrate. Women tend to engage in such activities and support the small-scale fishing activities that are undertaken mainly by men. Hence, small-scale fishing continues, regardless of the availability of fish, thus indicating the identity and resilience of this way of life. Furthermore, both the Indian and the Sri Lankan case studies show that some fishers are opting to stay in fisheries, either asserting their tradition and pride in their occupation and skills or because the alternatives are even worse; others are leaving; and still others want their children to leave. However, many fishers are staying, despite claiming that they are leaving or wanting to leave.

Chapter 4, by Nadine Vanniasinkam, Mohamed Faslan, and Nireka Weeratunge, deals with postwar conflicts over fish resources between migrant communities on the west coast and host communities on the east coast of Sri Lanka, in the context of the depletion of coastal resources and environmental changes. Discourses on 'the right to migrate' to pursue a mobile resource and 'the right to one's own resources' have emerged between migrant and host fisher groups that are contesting the seasonal access to fish on the east coast. The chapter explores horizontal relations between migrant and host fisher groups and vertical relations between fishers and the state, in order to identify the political-economic factors underlying emerging tendencies to ethnicize or regionalize these conflicts.

To a large extent, Asian coastal land is being exploited for tourism, harbours, infrastructure, and industry (Hall et al., 2011). Frequently, especially since the 2004 tsunami, fisherfolk are being pushed out of their traditional villages due to land grabbing, and they are finding it too difficult to continue fishing and fish-related activities. While many boat owners have benefited greatly from the developments, others who owned traditional fishing vessels lost their access to the sea and lagoons and hence their previous livelihoods. In particular, young people have increasingly migrated to other states and even abroad. These moves are also gendered: young men are moving to work on large fishing vessels, and young women are leaving fishing to take up work in urban areas and abroad (as maids and in industry, such as described Chapter 2, on Tamil Nadu). As a result, elderly women and the women left behind in traditional villages are facing a precarious work situation and are often the poorest of the poor. For example, in Chapter 5, Ragnhild Lund and Fazeeha Azmi show how coastal areas in eastern and northeastern Sri Lanka are increasingly being converted from fishing and fish landing sites to tourist resorts, industrial sites, and a harbour. These changes are the result of the war and displacement, the 2004 tsunami and resettlement, and development projects. Lund and Azmi's chapter addresses how those who are most vulnerable and excluded in these areas, namely female heads of household and their families, remain involved in fish vending or influence other people's choice to migrate abroad instead of engaging in fishing and living in a precarious situation. Another example is Raksa Sok and Kyoko Kusakabe's study of the role of ethnic Vietnamese in Cambodia and their inability to move freely (Chapter 8). Even when they are born and brought up in Cambodia, they are not able to obtain Cambodian citizenship, which prevents them from accessing education and deprives them of the right to own property. The decrease in fish resources has hit the ethnic Vietnamese hard. Some live on their savings, and others have started businesses and become involved in aquaculture. The differences in their responses to their situation have influenced their gender roles and household relations.

## *Gender*

In our project, gender was seen as a social construction, which means that men and women were socialized into different gender roles depending on the contextual and sociocultural factors at a place. The perspective captured differences in gender roles and functions among the specific countries and local communities. Whereas women were once central to fishing, they are now seen as homemakers, with responsibility for the care of the family and reproduction of the household, especially when the men are away. Several of the chapters in this book study how gender relations are changing now that women tend to be less involved in fishing (Chapters 2, 5, and 8). The exception to this is fish vending, which continues to be practised by middle-aged and elderly women (Chapter 5). Another trend is that married women are becoming increasingly tied to domestic work in the home through marriage (Chapter 2) and/or taking jobs outside the fisheries sector (Chapters 3, 4, 5, 6, and 8).

Furthermore, as this book shows, young men are taking up work as fish labourers on big boats and are commuting from their homes to fish for longer periods of time and over greater distances. Men are also migrating from outside fishing communities to take up work on boats. The individual chapters provide further insights into such issues.

### *Vulnerability and precarity*

The chapters in this book deal with vulnerability and precarity issues in different ways, such as the impacts of the mechanization of fisheries (Chapter 2), coping strategies (Chapter 3), social exclusion (Chapters 5 and 8), conflict (Chapter 4), and immobility (Chapter 8). The fisheries sectors in all of the study countries are experiencing overfishing and growing social inequalities (Amarasinghe, 1989; Bavinck and Johnson, 2008; Salagrama, 2006). As documented in this book, the mechanization of fishing fleets and the introduction of other technologies have transformed the labour force into boat owners and labourers (Chapter 2). Such a replacement of traditional fishing practices with modern ones, dependent on capital-intensive technologies, continues to challenge the livelihoods of small-scale fishers. In India and Sri Lanka, these changes were accelerated and intensified by the post-tsunami reconstruction programmes (Juran, 2012). Through the programmes, the extensive distribution of boats and nets resulted in growing competition and increased capitalization, creating a new 'middle class' of boat-owning fishers (Chapters 2 and 6). Alternatively, they created conflict (Chapter 4) and the exclusion of fishers from fishing (Chapter 5).

### *Well-being and identity*

Various chapters in this book study the links between mobility and migration on the one hand and well-being on the other hand. Chapter 6, by Nireka Weeratunge, Ramani Gunatilaka, Nadine Vanniasinkam, Mohamed Faslan, Dilanthi Koralagama, and Nirmi Vitarana, addresses how households in fishing communities on the west and east coasts of Sri Lanka engage in internal, coast-to-coast seasonal migration to pursue fishing and fisheries-related livelihoods. The chapter focuses on decision-making within households and argues for a gendered understanding of the material, relational, and subjective motivations underlying the pursuit of well-being by women and men as driving migration, rather than poverty or resource scarcity. Furthermore, in Chapter 7, Nitya Rao and C.M. Pratheepa find that several factors shape Indian fishers' decisions to fish in distant waters. Apart from resource depletion, growing competition, and changes in fishing technologies and practices as a push factor, many of the fishers identify as specialists in deep-sea fishing, especially as 'shark hunters'. More than earning money, they gain a sense of satisfaction from displaying their skills and from confronting and overcoming serious risks.

Chapter 12, by Ramani Gunatilaka, adds value to the individual-country studies on well-being and migration by providing a cross-country analysis of migration and consumption patterns. She documents how the economics literature on fishing communities in developing countries is concerned mainly with poverty and access to resources. However, in the chapter, Gunatilaka uses survey-based data from inland and coastal fishing communities in Cambodia, India, and Sri Lanka to study the impact of migration and movement in multiday boats on the distribution of fishing households' consumption expenditure. The situation in the three countries is shown to vary geographically and economically in relation to the effect of migration on poverty and the distribution of consumption.

### *Networks and collective action*

This book includes three chapters that investigate how collective action and networks work for or against the well-being of small-scale fishers in Cambodia, India, and Sri Lanka. The chapters present different types of networks that attempt to generate well-being and overcome vulnerability and 'precarity', including institutional networks in Cambodia and semi-informal collective actions through fishing associations in Sri Lanka and informal actions in India.

Two chapters (Chapters 9 and 10) focus on forms of social capital that have been identified as bonding, bridging, and linking, which describe different kinds of interpersonal relationships (Woolcock and Narayan, 2000). Bonding social capital describes horizontal relationships between individuals who identify as part of the same group (e.g. family and caste), and bridging social capital describes relationships between individuals who identify as members of different groups (villages and religion). Linking capital describes vertical relationships between different groups with different access to power, such as between government employees and rural villagers or between donors and recipients (Woolcock and Narayan, 2000). The types of social capital may have either strong or weak ties, thus bonding social capital is not synonymous with strong ties, whereas bridging may consist of strong or weak ties (Leonard and Onyx, 2003).

In Chapter 9, Nitya Rao, R. Manimohan, and C.M. Pratheepa study the different types of migration in coastal Tamil Nadu and the role of kinship and other support networks in enabling such movement. They unpack the differences between various migrant streams: overseas migration for fishing and non-fishing, migration for multiday fishing, and distress migration from other parts of India. In the context of precarious labour, long-established kin and gender networks in Kanyakumari District contribute to relatively positive experiences of migration and well-being outcomes relative to Cuddalore District. In Chapter 10, Nirmi Vitarana, Dilanthi Koralagama, Nireka Weeratunge, and Ramani Gunatilaka explore the importance of social relations and networks for enabling or constraining migration in Sri Lankan fishing communities. They examine how different types of networks provide support for women's and men's livelihoods, including migration, as well as the comparative strengths and weaknesses of social networks in communities in two districts: Puttalam District and Trincomalee District. They discuss how social networks are used by informal and formal groups to mobilize collective action for and against migration, and they argue that stronger bridging and linking ties in west coast fishing communities help to sustain migration. By contrast, east coast communities have weaker networks and thus reveal social exclusion.

Chapter 11, by Kyoko Kusakabe, Prak Sereyvath, Lam Doeurn, and Yem Sivon, addresses the Fisheries Administration's establishment of community fisheries (CFs) in Cambodia. CFs are given the right to manage and regulate fishing resources in designated fishing areas. However, attention has recently turned to contestations about the coastal land resource. The shift in focus has also helped the shift in leadership, as women are now taking up more leadership positions than they were before. The authors analyse cases of two CFs and their struggles against land grabbing.

## Methodology

The qualitative data (interviews and focus group discussions) and quantitative data (survey) used in the research constitute the basis for this book (Table 1.1). This enabled us to understand key issues that facilitated cross-country comparisons and enhanced our understanding of what was unique and country-specific. The quantitative information complemented the information gained from qualitative methods. While the ethnography detailed *why* and *how* mobile livelihoods and gendered subjectivities differ in fishing communities, the surveys provided detailed information on the *patterns* of the processes uncovered through the ethnographic data. Our combination of qualitative and quantitative methodologies was an unusual feature for a study of migration in small fishing communities. Pilot projects in each country involved local participation and interventions that directly served

*Table 1.1* Overview of the studies in the project Migration and Collectives/Networks as Pathways out of Poverty: Gendered Vulnerabilities and Capabilities amongst Poor Fishing Communities in Asia

| *Country* | *Cambodia* | *India* | *Sri Lanka* | *Sri Lanka* |
|---|---|---|---|---|
| **Study location** | Mekong (2 CFs*), coast (5 CFs), Tonle Sap (4 CFs), and Tonle Sap flooded area (3 CFs) | Cuddalore District and Kanyakumari District, 6 villages | Puttalam District and Trincomalee District, 4 villages | Batticaloa District and Ampara District, Batticaloa town and 3 villages |
| **Qualitative** | | | | |
| Interview | Key informant (19)<br>Open-ended (78 men and 144 women) | Key informant (24)<br>Semi-structured (50 men and 50 women)<br>Open-ended/life history (20) | Key informant (24)<br>Semi-structured/ livelihood profiles (53)<br>Open-ended/life history (44 women and 39 men) | Key informant (5)<br>Open-ended/life history (7 women) |
| Group discussions | Separate FGDs* with men (16), women (16), and youths (14) | Separate FGDs with men (12), women (12), and youths (12) | Separate FGDs with men (4), women (4), and youths (6) | Men and women, old and young mixed groups (4) |
| Observation | Process documentation | Field observations | Field notes | Field notes |
| **Quantitative** | | | | |
| Survey | Socio-Economic Survey 2011 (3500 households) | 633 households, 2017 (men and women separately) | 800 households, 2017 (men and women separately) | – |
| **Pilot projects** | | | | |
| Participatory and/or action oriented | Process documentation, ecotourism video | Video documentation on precarity, participatory community discussions, and documentation | Participatory videos by youths on livelihoods, migration and aspirations; and process documentation | – |

*FGDs: focus group discussions
Notes: CFs – Community Fisheries
Source: Data collected for the project 2017–2019

the communities and people involved in the research. The methodological triangulation facilitated knowledge production at different scales and across the countries, even though they both were contextually different. Data collection and fieldwork were conducted in the respective countries in the years 2016–2018.

The literature on small-scale fishers lacks quantitative information about the study countries, particularly with respect to small-scale fisheries, as the sampling frames based on population censuses are designed to make the surveys nationally representative according to spatial criteria rather than according to characteristics such as occupation. Therefore, nationally representative sample surveys do not always have sufficient data about fishing households and characteristics specific to them, which are needed for the kind of analysis that we aimed to carry out in our study. To unmask such information, surveys of households in all three countries, drawing on an enumerator-administered questionnaire and national databases, focused on the significance of migration as a strategy to alleviate poverty. In the individual countries, multiple methods were conducted with primary research participants (male fishers and female fishers, key informants, and youths), including open-ended and/or semi-structured interviews, life histories, social network mapping, and focus group discussions. Life histories provided multidimensional perceptions of poverty and well-being, and they enabled us to understand the impacts of migration, gendered decision-making patterns and dynamics of change among individuals, households, and communities at critical turning points. Social network mapping allowed us to identify actors who facilitated or obstructed migration and other livelihood strategies. Focus group discussions and key informant interviews gave us insights into the multidimensional perceptions of poverty, deprivation, and well-being; the impacts of migration; and gendered decision-making patterns.

The main fieldwork sites were selected such that the researchers worked in countries well known to them. The village sites were selected to create geographical variation (inland and marine fishing), variety in fishing types (high mechanized fishing overseas and artisanal and motorized boats), and different types of migration (overseas, seasonal, and interstate migration; distress migration; and displacement). The fieldwork sites are shown in Figures 1.1–1.3.

The methodology of the umbrella project aimed at working towards common goals. To that end, a two-day inception and methodology workshop was held in 2016 at the International Centre for Ethnic Studies (ICES), Sri Lanka, at the beginning of the project, where the main aims, methods, key concepts, and timelines were discussed. The Fisheries Management Resource Centre (FishMarc) in Puducherry organized a progress workshop in March 2018, at which the preliminary findings and dissemination strategies were discussed. In 2017 and 2018, the researchers attended a conference on 'precarity' at ICES and later took an active role in two fisheries

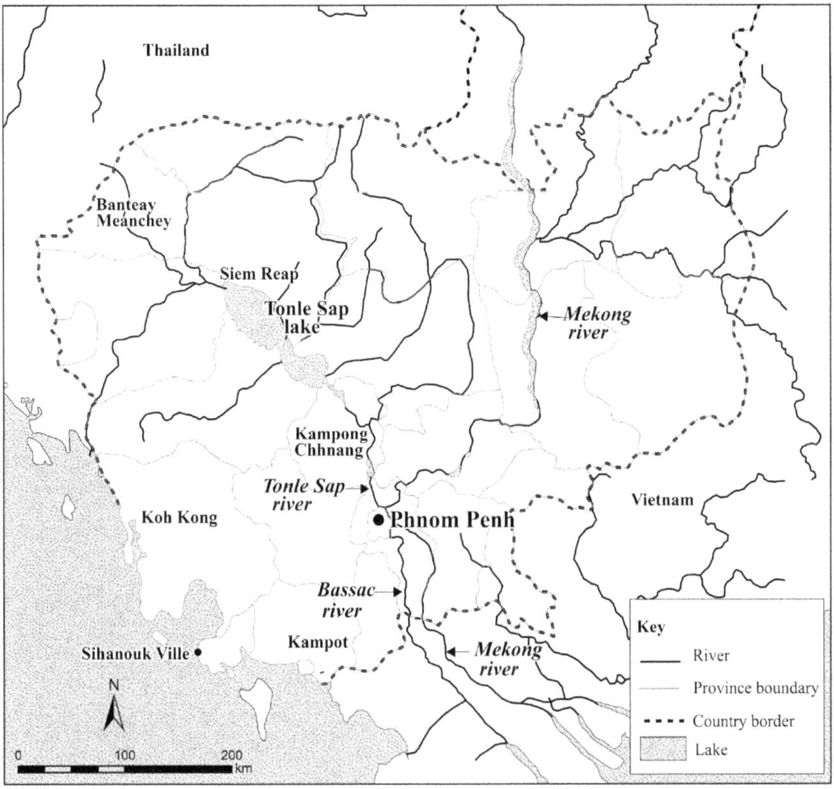


*Figure 1.1* Cambodia (inland and marine fisheries) – Tonle Sap lake and southwestern coast

Source: K. N. P. Piyatissa, University of Peradeniya

conferences and presented their work in special sessions (the 3rd World Small-Scale Fisheries Congress held in 2018 in Chiang Mai, and MARE, People of the Sea conference held in 2019 in Amsterdam) and individually (at the GAF7 Women and Aquaculture & Fisheries conference held in 2018 in Bangkok).

## Structure of the book

In the chapters to follow, we use information gained from the project to draw conclusions about the pressure on small-scale fishing and how it has dramatically changed since the late 2000s. The book is organized into four thematic sections, preceded by an introduction (Chapter 1) and followed by a conclusion (Chapter 13). Part I deals with how migration is related to

*Figure 1.2* India (marine fisheries) – Cuddalore District and Kanyakumari District, Tamil Nadu

Source: K. N. P. Piyatissa, University of Peradeniya

vulnerability, precarity, and social exclusion. Part II deals with social well-being and identity. Part III presents social relations, networks, and collective action. Part IV presents comparative perspectives on the distribution of consumption in fishing communities. All four sections present analyses from the three study countries. Although the different contributors work across different disciplinary and geographical boundaries, they position themselves actively towards the initial objectives of the umbrella research programme and engage actively with the conceptual keywords of mobility and migration, gender, vulnerability and precarity, well-being, and identity. Although there are multiple ways of engaging with these concepts, the different chapters elucidate that there is no longer one fixed and independent

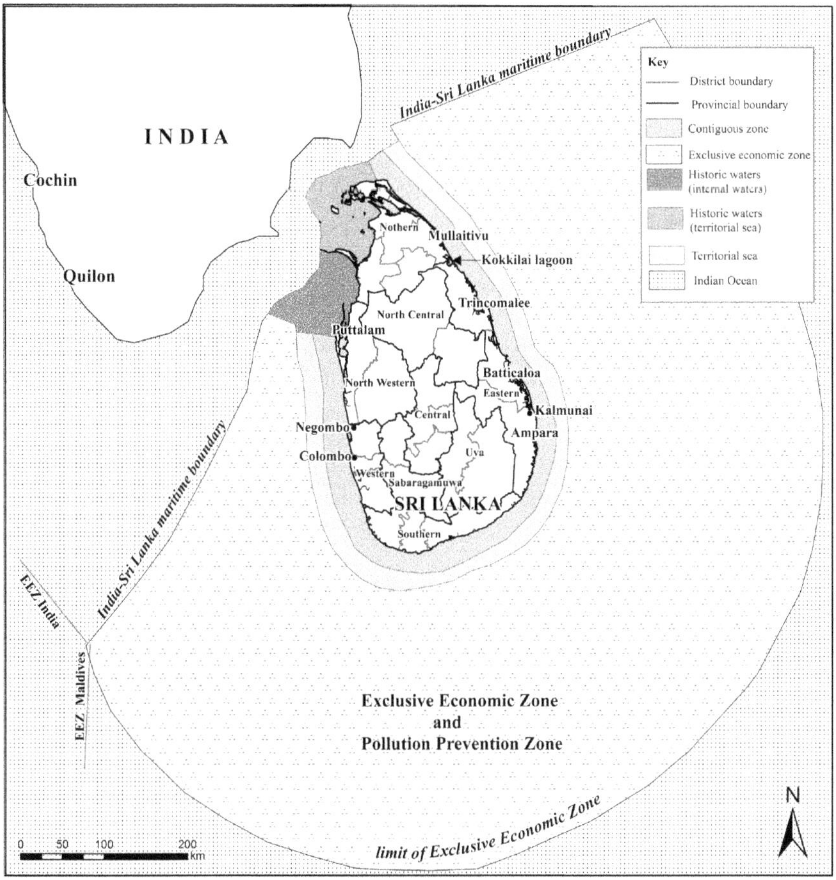


*Figure 1.3* Sri Lanka (marine fisheries) – Puttalam District, Trincomalee District, Batticaloa District, and Ampara District

Source: K. N. P. Piyatissa, University of Peradeniya

way of viewing the links between migration, vulnerability, and well-being. Rather, the researchers have engaged with the concepts and realities on the ground in different ways, leading us to come to grips with contextually different realities, which reflect the high degrees of variation within Asia.

## Notes

1 P. Zacharia, A. Gopalakrishnan, G. Gopakumar, M. Muralidhar, and K. Vijayan, 'Climate change impact on coastal fisheries and aquaculture in the SAARC region', paper presented at SAARC Agriculture Centre Video Conference on Climate Change Impact on Coastal Fisheries and Aquaculture, 20 December 2016

2 J. Allister McGregor, 'Human wellbeing in fishing communities', paper presented at the ESPA Workshop 1, held at the Institute of Development Studies, 2009

## References


Abobi, S.M., and Alhassan, E.H. (2015). A review of fisheries-related human migration in the Gulf of Guinea. *Journal of Coastal Zone Management, 18*. Retrieved from www.longdom.org/abstract/a-review-of-fisheriesrelated-human-migration-in-the-gulf-of-guinea-10747.html accessed 19/12/2019.

Allison, E.H., and Ellis, F. (2001). The livelihoods approach and management of small-scale fisheries. *Marine Policy, 25*(5), 377–388. Retrieved from https://doi.org/10.1016/S0308-597X(01)00023-9 accessed 19/12/2019.

Allison, E.H., Perry, A.L., Badjeck, M-C., Adger, W.N., Brown, K., Conway, D., . . . Dulvy, N.K. (2009). Vulnerability of national economies to the impacts of climate change on fisheries. *Fish and Fisheries, 10*(2), 173–196. Retrieved from https://doi.org/10.1111/j.1467-2979.2008.00310.x accessed 19/12/2019.

Amarasinghe, O. (1989). Technical change, transformation of risks and patronage relations in a fishing community of South Sri Lanka. *Development and Change, 20*(4), 701–733.

Aswathy, P., and Kalpana, K. (2018). The 'stigma' of paid work: Capital, state, patriarchy and women fish workers in South India. *Journal of International Women's Studies, 19*(5), 113–128.

Athulathmudali, S., Balasuriya, A., and Fernando, K. (2011). *An exploratory study on adapting to climate change in coastal areas of Sri Lanka*. Working Paper Series No. 18 Colombo: Centre for Poverty Analysis.

Badjeck, M.C., Allison, E.H., Halls, A.S., and Dulvy, N.K. (2010). Impacts of climate variability and change on fishery-based livelihoods. *Marine Policy, 34*(3), 375–383. Retrieved from https://doi.org/10.1016/j.marpol.2009.08.007 accessed 19/12/2019.

Bavinck, M., and Johnson, D. (2008). Handling the legacy of the blue revolution in India: Social justice and small-scale fisheries in a negative growth scenario. *American Fisheries Society Symposium, 49*(1), 585–599.

Bavinck, M., and Vivekenandan, V. (2010). Conservation, conflict and the governance of fisher wellbeing: Analysis of the establishment of the Gulf of Mannar National Park and Biosphere Reserve. *Environmental Management, 47*, 593–602.

Béné, C. (2003). When fishery rhymes with poverty: A first step beyond the old paradigm on poverty in small-scale fisheries. *World Development, 31*(6), 949–975. Retrieved from https://doi.org/10.1016/S0305-750X(03)00045-7 accessed 19/12/2019.

Béné, C., Al-Hassan, R.M., Amarasinghe, O., Fong, P., Ocran, J., Onumah, E., . . . Mills, D.J. (2016). Is resilience socially constructed? Empirical evidence from Fiji, Ghana, Sri Lanka, and Vietnam. *Global Environmental Change, 38*, 153–170.

Betcherman, G., and Marschke, M. (2016). Coastal livelihoods in transition: How are Vietnamese households responding to changes in the fisheries and in the economy? *Journal of Rural Studies, 45*, 2433. doi:10.1016/j.jrurstud.2016.02.012

Blaikie, P. (1999). A review of political ecology: Issues, epistemology and analytical narratives. *Zeitschrift fur Wirtschaftsgeographie, 43*(3–4), 131–147.

Bourdieu, P. (1977). *Outline of a theory of practice* (R. Nice, Trans.). Retrieved from https://doi.org/10.1017/CBO9780511812507 accessed 19/12/2019.

Britton, E. (2012). Women as agents of wellbeing in Northern Ireland's fishing households. *Maritime Studies*, *11*. Retrieved from https://doi.org/10.1186/2212-9790-11-16 accessed 19/12/2019.

Britton, E., and Coulthard, S. (2013). Assessing the social wellbeing of Northern Ireland's fishing society using a three-dimensional approach. *Marine Policy*, *37*, 28–36. Retrieved from https://doi.org/10.1016/j.marpol.2012.04.011 accessed 19/12/2019.

Coulthard, S. (2015). Can we be resilient and well, and what choices do people have? Incorporating agency into the resilience debate from a fisheries perspective. *Ecology and Society*, *17*(1), 4. Retrieved from http://dx.doi.org/10.5751/ES-04483-170104 accessed 19/12/2019.

Coulthard, S., Johnson, D., and McGregor, J.A. (2011). Poverty, sustainability and human wellbeing: A social wellbeing approach to the global fisheries crisis. *Global Environmental Change*, *21*, 453–463.

Coulthard, S., Sandaruwan, L., Paranamana, N., and Koralgama, D. (2014). Taking a well-being approach to fisheries research: Insights from a Sri Lankan fishing village and relevance for sustainable fisheries. In L. Camfield (Ed.), *Methodological challenges and new approaches to research in international development* (pp. 76–100). Basingstoke: Palgrave Macmillan.

Crona, B., and Rosendo, S. (2011). Outside the law? Analyzing policy gaps in addressing fishers' migration in East Africa. *Marine Policy*, *35*(3), 379–388.

D'Anna, L.M., and Murray, G.D. (2015). Perceptions of shellfish aquaculture in British Columbia and implications for well-being in marine social-ecological systems. *Ecology and Society*, *20*(1), 57. Retrieved from http://dx.doi.org/10.5751/ES-07319-200157 accessed 19/12/2019.

De Silva, D.A., and Yamao, M. (2007). Effects of the tsunami on fisheries and coastal livelihood: A case study of tsunami-ravaged southern Sri Lanka. *Disasters*, *31*(4), 386–404.

Department of Animal Husbandry Dairying & Fisheries. (2017). *Fisheries profile of India*. Retrieved from www.dahd.nic.in/about-us/divisions/fisheries accessed 5/11/18.

FAO. (2019a). *Securing sustainable small-scale fisheries: Sharing good practices from around the world*. FAO Fisheries and Aquaculture Technical Paper 644. Retrieved from www.fao.org/3/CA3041EN/ca3041en.pdf accessed 19/12/2019.

FAO. (2019b). *Small-scale fisheries and aquaculture & family farming*. Retrieved from www.fao.org/family-farming/themes/small-scale-fisheries/en/

Ferguson, J. (2015). *Give a man a fish: Reflections on the new politics of distribution*. , NC: Duke University Press.

Food and Agricultural Organization of the United Nations. (2015). *Voluntary guidelines for securing sustainable small-scale fisheries in the context of food security and poverty eradication*. Rome: Food and Agricultural Organization of the United Nations.

Gopal, N., Williams, M.J., Porter, M., Kusakabe, K., and Choo, P.S. (2014). Gender in aquaculture and fisheries: Navigating change. *Asian Fisheries Society*, *27S*, 1–14.

Hall, D., Hirsch, P., and Murray Li, T. (2011). *Powers of exclusion: Land dilemmas in Southeast Asia*. Honolulu, HI: University of Hawai'i Press.

Hapke, H.M. (2001). Gender, work, and household survival in South Indian fishing communities: A preliminary analysis. *Professional Geographer*, *53*(3), 313–331.

Hapke, H.M. (2017). Theorizing patriarchy: Development paradoxes and the geography of gender in South Asia. *Gender, Technology and Development*, *17*(1), 1–29.

Hapke, H.M., and Ayyankeril, D. (2018a). Gendered livelihoods in the global fish-food economy: A comparative study of three fisherfolk communities in Kerala, India. *Maritime Studies*, *17*, 133–143.

Hapke, H.M., and Ayyankeril, D. (2018b). Gulf migration and changing patterns of gender identities in a South Indian Muslim community. In M. Amrith and N. Sahraoui (Eds.), *Work and migration: Agency in a gendered labour setting* (pp. 175–192). Abingdon: Routledge.

Heinonen, U. (2006). Environmental impact on migration in Cambodia: Water-related migration from the Tonle Sap Lake region. *International Journal of Water Resources Development*, *22*(3), 449–462. Retrieved from https://doi.org/10.1080/07900620500482865 accessed 19/12/2019.

Islam, M.M., and Herbeck, J. (2013). Migration and translocal livelihoods of coastal small-scale fishers in Bangladesh. *Journal of Development Studies*, *49*(6), 832–845. doi:10.1080/00220388.2013.766719 accessed 19/12/2019

Jeeva, J.C., Balasubramaniam, S., Jeyanthi, P., and Ashaletha, S. (2011). Evaluation of the post-tsunami scenario with reference to fishing technology and socioeconomic conditions among the motorized craft operators in Tamil Nadu. *Indian Journal of Fisheries*, *58*(3), 117–123. Retrieved from http://drs.cift.res.in/handle/123456789/695 accessed 19/12/2019.

Johnson, D.S., Acott, T.G., Stacey, N., and Urquhart, J. (Eds.). (2018). *Social wellbeing and the values of small-scale fisheries*. Cham: Springer International.

Juran, L. (2012). The gendered nature of disasters: Women survivors in post-tsunami Tamil Nadu. *Indian Journal of Gender Studies*, *19*(1), 1–29. Retrieved from https://doi.org/10.1177/097152151101900101 accessed 19/12/2019.

Kadfak, A., Bennett, N., and Prugsamatz, R. (2012). *Scoping study on migrant fishers and transboundary fishing in the Bay of Bengal*. Phuket: Bay of Bengal Large Marine Ecosystem Project (BOBLME). Retrieved from http://aquaticcommons.org/19036/

Kraan, M.L. (2009). *Creating space for fishermen's livelihoods: Anlo-Ewe beach seine fisher's negotiations for livelihood space with multiple governance structures in Ghana*. African Studies Collection 19. Leiden: African Studies Centre.

Kramer, R.A., Simanjuntak, S.M.H., and Liese, C. (2002). Migration and fishing in Indonesian coastal villages. *Ambio*, *31*(4), 367–372.

Krishnan, V. (2010). *An economic study on migration of fishermen from Kanyakumari District*. Doctoral thesis, Manonmaniam Sundaranar University, Tirunelveli.

Kusakabe, K. (2014). *Women in fish value chain and collective business: Case of processed fish traders in Cambodia*. Paper presented at the 5th Global Symposium on Gender in Aquaculture and Fisheries (GAF5) Special Symposium at the 10th GAF Forum, Lucknow, India, 13–15 November 2014.

Kusakabe, K. (2017). Analysis 2: SDG5: Gender equality in and through fisheries. *Yemaya*, *54*, 5–6.

Leonard, R., and Onyx, J. (2003). Networking through loose and strong ties: An Australian qualitative study. *Voluntas: International Journal of Voluntary and Non-Profit Organisations*, *14*(2), 189–203.

Marquette, C.M., Koranteng, K.A., Overå, R., and Aryeetey, E.B-D. (2002). Small-scale fisheries, population dynamics, and resource use in Africa: The case of Moree, Ghana. *AMBIO*, *31*(4), 324–336.

Mosse, D. (2010). A relational approach to durable poverty, inequality and power. *Journal of Development Studies*, *46*(7), 1156–1178. Retrieved from https://doi.org/10.1080/00220388.2010.487095 accessed 19/12/2019.

Murray Li, T. (2011). Centering labor in the global land grab debate. *Journal of Peasant Studies*, *38*(2), 281–298.

Murray Li, T. (2014). Involution's dynamic other. *Journal of Royal Anthropological Institute*, *20*(2), 276–292.

Murray Li, T. (2017). After the land grab: Infrastructural violence and the 'mafia system' in Indonesia's oil palm plantation zones. *Geoforum*, *96*, 328–337. doi:10.1016/j.geoforum.2017.10.012

Njock, J-C., and Westlund, L. (2010). Migration, resource management and global change: Experiences from fishing communities in West and Central Africa. *Marine Policy*, *34*(4), 752–760. Retrieved from http://dx.doi.org/10.1016/j.marpol.2010.01.020 accessed 19/12/2019.

Nuorteva, P., Keskinen, M., and Varis, O. (2010). Water, livelihoods and climate change adaptation in the Tonle Sap Lake area, Cambodia: Learning from the past to understand the future. *Journal of Water and Climate Change*, *1*(1), 87–101. Retrieved from https://doi.org/10.2166/wcc.2010.010

Overå, R. (2001). *Institutions, mobility and resilience in Fante migratory fisheries in West Africa*. CMI Working Paper WP 2001:2. Bergen: Chr. Michelsen Institute.

Overå, R. (2005). Institutions, mobility and resilience in the Fante migratory fisheries in West Africa. *Transactions of the Historical Society of Ghana*, *9*, 103–123.

Ram, K. (1991). *Mukkuvar women: Gender, hegemony and capitalist transformation in a south Indian fishing community*. London: Zed Books.

Rigg, J., Oven, K.J., Basyal, G.K., and Lamichhane, R. (2016). Between a rock and a hard place: Vulnerability and precarity in rural Nepal. *Geoforum*, *76*, 63–74.

Salagrama, V. (2006). *Trends in poverty and livelihoods in coastal fishing communities of Orissa State, India*. FAO Fisheries Technical Paper 490. Retrieved from www.fao.org/3/a-a0692e.pdf accessed 19/12/2019.

Sen, A. (1999). *Development as freedom*. Oxford: Oxford University Press.

Sneddon, C. (2007). Nature's materiality and the circuitous paths of accumulation: Dispossession of freshwater fisheries in Cambodia. *Antipode*, *39*(1), 167–193. Retrieved from https://doi.org/10.1111/j.1467-8330.2007.00511.x accessed 19/12/2019.

Trimble, M., and Johnson, D. (2013). Artisanal fishing as an undesirable way of life? The implications for governance of fishers' wellbeing aspirations in coastal Uruguay and southeastern Brazil. *Marine Policy*, *37*, 37–44.

United Nations. (2019). *Sustainable development goals*. Retrieved from https://unstats.un.org/sdgs accessed 19/12/2019.

Vivekanandan, V., and Kasim, M. (2011). *Fisheries management options for Tamil Nadu & Puducherry*. Retrieved from www.academia.edu/33959830/Fisheries_Management_Options_for_Tamil_Nadu_and_Puducherry accessed 19/12/2019.

Weeratunge, N., Béné, C., Siriwardane, R., Charles, A., Johnson, D., Allison, E.E., and Badjeck, M-C. (2014). Small-scale fisheries through the wellbeing lens. *Fish and Fisheries*, *15*(2), 255–279. Retrieved from https://doi.org/10.1111/faf.12016 accessed 19/12/2019.

Weeratunge, N., Snyder, K.A., and Sze, C.P. (2010). Gleaner, fisher, trader, processor: Understanding gendered employment in fisheries and aquaculture. *Fish and Fisheries*, *11*(4), 405–420.

White, S.C. (2008). *But what is wellbeing? A framework for analysis in social and development policy and practice*. Unpublished manuscript, University of Bradford. Retrieved from http://people.bath.ac.uk/ecsscw/But_what_is_Wellbeing.pdf accessed 19/12/2019.

White, S.C. (2017). Relational well-being: Re-centering the politics of happiness, policy and the self. *Policy & Politics*, *45*(2), 121–136.

Wijayaratne, B., and Maldeniya, R. (2003). *The role of fisheries sector in the coastal fishing communities of Sri Lanka*. Retrieved from http://pubs.iclarm.net/resource_centre/AMF_Chapter-26-FA.pdf accessed 19/12/2019.

Woolcock, M., and Narayan, D. (2000). Social capital: Implications for development theory, research, and policy. *World Bank Research Observer*, *15*(2), 225–249.

# 2 Migration for capital accumulation

## Changing class dynamics among small-scale fishers on the Coromandel Coast, Tamil Nadu

*Nitya Rao and R. Manimohan*

### Introduction

India has witnessed a gradual mechanization of marine fisheries since the 1960s, popularly called the Blue Revolution (Bavinck, 2001). The introduction and expansion of trawling coincided with the location of rich shrimp grounds in India's Exclusive Economic Zone and the development of an export-oriented shrimp industry (Pillai and Katiha, 2004).[1] The impetus for this shift has been provided by state investments in infrastructure such as boatyards, harbours, and post-harvest technologies; the establishment of the Marine Products Export Development Authority (in 1972); and subsidies for fishing vessels, fishing gear, and fuel. The marine fisheries focus on commercialization and export-oriented reforms, reinforced by the National Policy on Marine Fisheries, 2017 (Government of India, 2017), has attracted private sector investments in several support industries, leading India to rank seventh in global marine capture fish production (Sathianandan, 2017).

In India, marine fishing is largely a caste-based occupation. However, the introduction of mechanization marked the beginning of visible class formation among the fishers, between trawler owners, small-scale fishers, and working-class employees on the trawlers and in the service sector in the vicinity of the harbours (Bavinck, 2014; Ram, 1991; Vivekanandan and Kasim, 2011). The most recent statistics show that 34% of the marine fishers in India work in the mechanized fishing sector and account for 70% of the catch, which is twice the amount caught by those using motorized boats and nine times more than the amount caught by traditional fishers (Sathiadhas, 2009). Average incomes reflect this disparity between types of fishing vessels, with the annual per capita earnings of fishing labourers ranging from INR13,200 (USD186) for those working on a motorized dinghy with a bag net to INR127,200 (USD1790.5) for those working on a mechanized purse seiner (Sathiadhas, 2009). It is therefore not surprising that although marine fisheries in India have often been considered as offering better pay than other primary sectors in the country (Bavinck, 2014), almost 61% of fishing families are classified as below the poverty line (CMFRI, personal communication, 15 March 2018). These trends exist in Tamil Nadu.

The 2004 tsunami marked a turning point, as the debilitating losses of both lives and property led to the uncontrolled distribution of fishing equipment, which in turn led to almost all artisanal fishing off the east coast of Tamil Nadu being done from motorized vessels (Bavinck, 2008). This, together with the expansion of the trawling fleet, intensified the density of fishing vessels and competition within the artisanal fishing sector. In this chapter, which is based on data collected from Cuddalore District in Tamil Nadu, we explore how small-scale artisanal fishers are adapting to the decline in near-shore fish catches and the rapid mechanization of the sector. In particular, we address the following questions: To what extent are artisanal fishers' strategies – whether in terms of labour relations, including migration for work or the employment of in-migrants as crew, or in terms of institutional mechanisms, especially the sharing of boat ownership – enabling a more equitable distribution of resources and benefits? In what ways are their strategies further entrenching inequalities, including those relating to gender?

In the next section, we briefly discuss our conceptual starting points, followed by a description of the context and methods. Thereafter, we explore how capital accumulation is creating new forms of precarity. We focus particularly on the interlinkages between technology, capital, and labour and their implications for class-based, gendered, and generational subjectivities in fishing communities.

## Conceptual starting points

Research on economic growth and sustainable development, including in the field of marine or inland fisheries, has emerged from simultaneous concerns relating to the politics of resource depletion and environmental conservation, livelihood security in the face of unregulated privatization and new forms of accumulation, and social justice amid growing inequalities between 'winners' (those who benefit) and 'losers' (those who bear the costs) (Bavinck et al., 2018; Campling et al., 2012). Although efforts have been made to address 'old' or 'inherited' forms of vulnerability, they have produced new forms of 'precarity' (Rigg et al., 2016, p. 66) (for a fuller discussion, see Chapter 1, this volume). The effects of these changes are not uniform but rather differentiated by caste, gender, and generation and increasingly by class identities.

In addition to the need to understand the environmental changes and shifts in the political economy that drive changes in the small-scale marine fisheries sector, there is also a need to understand the cultural and technoscientific changes taking place in the communities involved in those fisheries. Consumption, when seen as a social communication system that extends beyond status to reflect people's life projects (Appadurai, 1986) and provide spaces for cultural assertion (Gidwani and Sivaramakrishnan, 2003), can provide insights into the often-circuitous and non-linear processes involved in accumulation (Sneddon, 2007). Moreover, outcomes often can be contradictory across spatial, temporal, and institutional scales. Intensive capital investment is pursued as a short-term strategy, providing immediate incomes and material gains, despite the realization of the implications of the longer-term non-viability and resource depletion of such an aggressive fishing strategy (Sundar, 2010). However, the desire to be a boat owner seems to override all other considerations, driving high levels of indebtedness and male migration in our two study villages in Cuddalore District. While there is no guarantee of 'success', women in their role as wives are willing to gamble with the only assets they have – their gold jewellery – to support the fulfilment of the aforementioned desire, thus demonstrating not just relations of interdependence but also their centrality in advancing collective life projects. The more subtle drivers, experiences, and outcomes of particular decisions can to some extent be exposed by gender analysis, which examines relationships of cooperation and conflict between and among men and women (Sen, 1990) and the power dynamics therein (Kandiyoti, 1998).

Migration and labour mobility are central to our analysis in this chapter given that they shape the possibilities for class formation among fishers: locals against migrants, small scale against large scale, and one group of fishers against another. In the context of small-scale fisheries in India, the growing movement of people is driven by a host of reasons: apart from the changing availability of fish species due to changes in ocean currents, temperature, and climate (Vivekanandan, 2011), they include differences in skills, knowledge and abilities, employment potential, and political and religious factors (Rajan, 2002, p. 6).

Fishers' mobility has both temporal and spatial dimensions and takes diverse forms, including daily trips, voyages for a few days or weeks, and longer-duration stays, often involving movement to a different destination (Cresswell and Merriman, 2011). Their migration overseas may be for fishing itself or for other forms of labour (as discussed in Chapter 9, this volume), with the objective of generating resources for investment in their own enterprises at home (Rao, 2013). At times, such movements are not voluntary or aspirational but driven by distress, as in the case of the in-migrants in the studied villages. A better understanding of migration and movement, its links to the new forms of fish capture and distribution, and how it is differently experienced by men and women can help to unravel the complexities of labour relations, income-sharing, and livelihood insecurity involved in the process.

## Context and methods

This chapter is based on a study of two villages in Cuddalore District, Periyapattinam and Uppanankuppam,[2] both of which were severely affected by the 2004 tsunami. The study comprised a household survey of 100 households, followed up by qualitative in-depth interviews with 10 households (husband

and wife) in each village. Focus group discussions were held with different categories of men and women to understand their perspectives and aspirations. This was important because the source of the current conflict over the use of ring seines appears partly generational, with educated young men not being able to find suitable employment and therefore seeking greater profits and incomes in fishing. Information on the challenges and opportunities in marine fisheries in the study villages was collected through key informant interviews with ring seine shareholders, crew members, female auctioneers, traders, representatives of NGOs, and in-migrants. The data were collected between March 2017 and February 2018.

Both Periyapattinam and Uppanankuppam are primarily fishing villages, dominated by the Pattinavar caste with two subcastes: Chinna Pattinavar and Periya Pattinavar. There is a small group of trawler owners in Periyapattinam, all of whom are Periya Pattinavars, but the majority are small-scale fishers, Chinna Pattinavars, currently using ring seine technologies. The former, who are also the traditional elite, control the traditional *ur panchayat* (village council), which is responsible for resource allocation, management, and conflict resolution and for organizing the annual temple festival (Bavinck and Vivekanandan, 2017). The Periya Pattinavars also control the harbour association, which is responsible for regulating the large number of people who are employed both directly and indirectly in the provision of goods and services, whether as ice providers, mechanics, cleaners, or hoteliers.

After the tsunami in 2004 and with the growing availability of the Indian oil sardine (*Sardinella longiceps*) off the east coast of India (Vivekanandan, 2011), male small-scale fishers in the neighbouring district of Nagapattinam, adopted the use of ring seines, a method of fishing that had been developed in Kerala in the 1980s. One of our key informants said,

> When the *surukuvalai* [ring seine] was used initially, we were against it. It was banned by our *ur panchayat* [village council]. In Pazhaiyaru [Nagapattinam District], they had started using these nets in *kanna* boats. Once they came close to our village for fishing. We protested, as the gear both destroys the sea bottom and intensifies the unequal distribution of income and wealth among us fishers. When they didn't stop, fishermen from our and neighbouring villages went in 200 boats to Pazhaiyaru [harbour in Tamil Nadu, to the south of Cuddalore District] and burnt their nets, yet the technology spread. We could not control it, but were losing out in terms of fish catches.
>
> (Manikkam, male, 53 years)[3]

Since the use of ring seines is destructive, the method is banned by the Tamil Nadu state government, yet the small-scale fishers in Periyapattinam (i.e. the Chinna Pattinavars), especially the young men, started arguing with the *ur panchayat* to lift the restrictions on its use. They were encouraged by traders because of the high demand for oil sardines and mackerel in Kerala. Despite the lack of a decision, they slowly started using ring seines on their motorized (10 HP) fibre-reinforced plastic (FRP) boats, which they had received as part of the tsunami relief efforts. When challenged about their control over capital and decision-making, both at the harbour and in the village, the *ur panchayat*, which was controlled by the trawler owners, convened a meeting and agreed to restrict ring seine operations to six months per year (Lawrence and Bhalla, 2018). Since the late 2000s, the numbers of ring seine units have rapidly increased in Cuddalore District, doubling in Periyapattinam from 56 in 2010 (CMFRI, personal communication, 15 March 2018) to 107 in 2017 (household survey 2017).

Uppanankuppam is smaller than Periyapattinam, and in 2010, it had four trawlers, 15 ring seine units, 215 motorized fishing vessels and 50 non-motorized fishing vessels (CMFRI, personal communication, 15 March 2018). Almost 60% of households in Uppanankuppam used their motorized FRP boats to secure shares in ring seine units, which contributed to the significant rise in ring seine shares noted in our survey (Table 2.1). While not necessarily in support of the ring seine technology, the households found they had no choice, and many complained that the biggest challenge they encountered was from the larger fishing villages, such as Periyapattinam, which put their livelihoods at risk. With growing competition, small-scale fishers were willing to make larger investments in steel boats, fitted with high-speed engines[4] and with sufficient storage and other facilities to enable longer voyages, whereby they hoped to increase profits and improve their living conditions.

Since the ring seine units have started to operate from harbours rather than from village beach landing sites, the change in the organization of fishing has had implications for gender relations, roles, and identities. The harbour located in the old town of Cuddalore has become the major business centre for both villages. Whereas men increasingly seek shares in boats and

*Table 2.1* Boat types owned by households in the study villages, Cuddalore District, Tamil Nadu, 2017

| *Village*[*] | *Rowing boats* | | *Motorized fibre-reinforced plastic boats* | | | *Mechanized boat* | | | | *Ring seine units* | |
|---|---|---|---|---|---|---|---|---|---|---|---|
| | *No. of HH*[**] | *No. of boats, own* | *HH* | *Own* | *Share* | *Own* | | *Share* | | *HH* | *No. of shares* |
| | | | | | | *HH* | *No.* | *HH* | *No.* | | |
| Periyapattinam | 15 | 15 | 40 | 40 | 7 | 4 | 4 | 1 | 1 | 59 | 62 |
| Uppanankuppam | 10 | 9 | 49 | 55 | 9 | 0 | 0 | 2 | 2 | 48 | 51 |
| Totals | 25 | 24 | 89 | 95 | 16 | 4 | 4 | 3 | 3 | 107 | 113 |

Source: Survey data, 2017; Notes: [*] village names are fictitious, [**] HH – households

*Table 2.2* Occupational profile of villagers in the study villages, Cuddalore District, Tamil Nadu, 2017

| *Occupation* | *Periyapattinam** | | *Uppanankuppam* | |
|---|---|---|---|---|
| | No. | % | No. | % |
| Male: | | | | |
| Fishing boat owner/shared ownership | 81 | 50.6 | 84 | 51.9 |
| Fishing crew/labourer | 27 | 16.9 | 30 | 18.5 |
| Fishing-related activity | 4 | 2.5 | 4 | 2.5 |
| Fish trader/ agent | 2 | 1.3 | 0 | 0.0 |
| Petty business | 3 | 1.9 | 3 | 1.9 |
| Private/state employee | 4 | 2.5 | 2 | 1.2 |
| Work abroad | 10 | 6.3 | 27 | 16.7 |
| Fishing boat owner/shared ownership and fishing crew/labourer | 29 | 18.1 | 12 | 7.4 |
| Totals | 160 | 100 | 162 | 100 |
| Female: | | | | |
| Fishing-related activity | 0 | 0.0 | 13 | 10.2 |
| Fish vendor (shop/marketplace) | 14 | 10.1 | 8 | 6.3 |
| Fish vendor (in another village) | 13 | 9.4 | 17 | 13.3 |
| Auctioneer | 0 | 0.0 | 9 | 7.0 |
| Petty business | 3 | 2.2 | 3 | 2.3 |
| Private/state employee | 9 | 6.5 | 0 | 0.0 |
| Household work | 100 | 71.9 | 78 | 61.0 |
| Totals | 139 | 100 | 128 | 100 |

Source: Survey data, 2017; Note: * village names are fictitious

simultaneously work as crew or at times migrate overseas to raise capital for shares, the possibilities open to women for fish vending and fishing-related activities appear to be limited. Approximately 20%–30% of women reported being engaged in fish-vending activities, while approximately 60–70% described themselves as homemakers (Table 2.2).

## Capital accumulation, precarity, and changing class relations

We start this section with extracts from a narrative by Raman – male, age 39, from Periyapattinam – to illustrate the social and economic processes involved in both gaining and challenging the elite control of fishing practices in the study villages:

> After the introduction of the ring seine, we got a lot of fish, oil sardines and mackerel. Initially, 16 families, all close relatives, came together to form a ring seine unit. We set a norm that each family contributed 20,000 rupees [USD287.20] as their share, so we could mobilize 320,000 rupees [USD4504]. We bought a boat with equipment from

Cuddalore harbour. A total of 30 members from our 16 families came as crew for fishing. We selected a person from among the shareholders to manage the day-to-day activities, such as account keeping, purchase of fuel and water, arranging finance, and equipment repair and maintenance. Everything went well for two years, but then problems arose and we stopped working together. Some of the shareholders questioned the trustworthiness of the accountant [*kanakkupillai*], but importantly people wanted to share the income only within a small group in order to accumulate wealth quickly. We sold the boat and shared the money among the partners. With others doing the same, today there are about 100 ring seine units in the village.

. . .

The current ring seine unit is my third one. The *kanna vallam* [plank-built canoe made of wood and fibre] was replaced by a steel boat with a powerful engine, [fishing] gear, GPS, and other equipment, almost twice in value. Now we are only five shareholders. The value of a share is 2 million rupees [USD28,152], 100 times what we first contributed, and altogether 10 million rupees [USD140,760] were mobilized from the shareholders. This time, I managed to buy only half a share, as I was unable to mobilize sufficient money. We have taken an advance of 2 million rupees from an auctioneer.

[. . .]

We have 40 crew members working on the boat. Most of them are small farmers and agricultural workers, in the age group 20 to 40 years, and they come from the agrarian hinterland. Some take advances from us to pay their children's school fees, for healthcare, or household consumption, so [they] are obliged to work on the boat until they repay the loan. We shareholders also work with the crew, but our tasks are more skilled, casting the net, observing the wind and water currents, and identifying the shoal (*maappu*), and are managerial in nature. We give instructions, which the workers follow.

[. . .]

With our steel boat, we can go for *thangal* (stay) fishing, staying at sea for about a week. Once we have our catch, we inform the auctioneer, so she can reach the harbour to auction our fish. The auctioneer gets 10% of the value of the catch as her commission for the advance given to the boat. During the ring seine season, there are always traders or their agents at the harbour to buy the catch.

[. . .]

We share the income on an equal basis. That is, after deducting the operational costs, including fuel costs of the main and carrier boats, food rations, water and ice, 50% goes to the owners and the remaining 50% goes to the crew. The crew includes the shareholders and the group of four to eight FRP boat operators, who work on carrier boats, transporting the catch to the harbour, in addition to their labour. As

> most of the shareholders are working partners in the boat, they get a share from the labour component of the income as well as their owner's share. The carrier boats are paid three labour shares [one share for the boat operator and two for the boat]. An extra half share is given to the boat driver, in addition to his labour share. Another half share is kept for giving incentives [payments] to efficient workers.
>
> [. . .]
>
> We do not share the income on a daily basis. Usually, once a month we check the accounts and calculate the income and expenditure and divide the income between owners and crew. If the catch and income is good, we may do this fortnightly. In case someone [owner or labourer] needs cash, he can get it as an advance from the accountant, to be deducted when the shares are calculated. The accountant is responsible for the money. Some may clandestinely lend for short-term interest, and this fuels doubt about the accountant's honesty.

The foregoing quote highlights a range of issues from the perspectives of both women and men relating to the mobilization of capital and labour and the social organization of ring seine fishing. In the following subsections, we discuss these issues and the role of migration and mobility in shaping them as processes, focusing specifically on the concepts of share and sharing.

### *Capital mobilization*

Raman's narrative reveals the need for a large amount of capital to set up a successful ring seine unit. With growing competition and declining catches per unit, fishers are seeking to establish partnerships and shareholding arrangements with smaller groups of people to enhance or at least maintain adequate earnings, yet in the process, they are producing new precarities for those without sufficient capital to buy shares. In their anxiousness to secure shares, excluded fishers adopt a range of strategies, collective and individual, some of which are riskier than others in terms of longer-term well-being and livelihood security.

At a collective level, the shareholders approach auctioneers – most of whom were women – for money to invest in their boats. Such investments give auctioneers the rights to auction the catches and earn commission. However, as investment values are increasing, auctioneers themselves often need to borrow from moneylenders. Jayathi, a 52-year-old female auctioneer, explained, 'In addition to my own resources, I borrowed Rs10 lakh [INR1 million, USD14,076] from various sources such as *seettu* [local chit fund] and private moneylenders. My co-auctioneer also bought her own share.'

At the individual level, in Periyapattinam and Uppanankuppam, there is a growing push towards labour migration as a strategy to accumulate capital to buy a share in a boat. Mama, a 70-year-old man from

Uppanankuppam, said, 'My younger son has been working in Singapore since 2011. From his earnings, he gave me 40,000 rupees [USD563] to buy a share in the ring seine here. He was very helpful and it is a good investment.'

Some of the younger male fishers have migrated to secure an income. Selvam went to Singapore in 2006 and worked there for three years to save enough money to buy a share in a ring seine unit. In the case of Satish, age 35, his brother worked in Singapore and sent money to Satish to buy a share in a unit. Satish also had to borrow a small amount of money from relatives. Many of these men are well educated, yet in the absence of other employment, they considered this the best route to livelihood security. Others, such as Sami, a 52-year-old man, considered themselves old to migrate overseas; instead, Sami worked as a labourer on a trawler and bought a share from the income.

Despite the contributions of the shareholding partners and the auctioneers, often from migrant remittances, the shareholders sometimes could not avoid borrowing additional amounts from moneylenders. The interest rates are as high as 60% per year, yet a number of the study participants mentioned that in 2009 the catches were good, incomes were high, and they could repay their loans after a few trips. However, this is no longer the case, with recent trips yielding either poor catches or no catches. Loans from private moneylenders are acquired against women's assets, specifically their gold jewellery, which they receive as dowry (Swaminathan et al., 2011):

> As ring seine fishing became popular and a money spinner, we wanted to buy a share [in a boat] too, but we did not have sufficient cash or savings. My husband had already started working on a ring seine boat as a crew member. We decided to raise capital by pledging my gold jewellery to a private finance company. While people say that fishermen can get . . . loans [against jewellery] from banks at low rates of interest, in practice it is difficult.
>
> (Shivani, female, age 37, from Periyapattinam)

The fishing sector is increasingly characterized as masculine, with much of the post-harvest work in the harbours being done by men (including migrant workers) rather than by women; this was the case when boats landed on the village beaches (for a detailed discussion of the social exclusion of women from fishing, see Chapter 5, this volume). Fewer opportunities for women to earn an income, combined with the need for more capital, has contributed to dowry inflation in the fishing communities, which in turn raises questions about women's 'declining' status, as reflected in the rapid decline in child sex ratios in Cuddalore District (Government of Tamil Nadu, 2017). Even for the auctioneers, unsustainable levels of debt have created new tensions and forms of precarity in their lives.

### Labour mobilization

Ring seine fishing requires a relatively large crew compared with trawling. Initially, labour was provided by kin, as mentioned by Raman, but as the number of ring seine units grew, there was a manifold increase in the demand for labour. Fishing villages were unable to meet the high demand, and therefore, boat owners started to recruit labour from the agrarian hinterlands. Because farmland in Cuddalore District is fed by rain and farming is seasonal, a section of the agrarian population had already moved to the coast from c. 2011 onwards, and Cuddalore old town harbour became an attractive centre for a number of shore-based and post-harvest activities, such as loading and unloading fish or working in ice factories, which provided employment during the dry season. Boat owners had day-to-day interactions with the shore workers, and the latter became the channel for the recruitment of crew. Although some of the shore workers shifted to ring seine fishing themselves, they had a significant role as informal agents who engaged more people from their farming villages to work on the shores. Some of the boat owners too visited the villages, and through their contacts there, they directly recruited young workers by giving advances (loans) and *beta* (allowances for travel and food). Of the 30–40 labourers required to operate a ring seine unit, at least half came from farming villages.

In addition, labourers come from farther afield, not just from other districts of Tamil Nadu. We found a large settlement of 300–400 migrant workers from northern India, near the two study villages. Although some of them worked in the local industrial complex, more than 150 Muslim labourers from Farrukhabad District in Uttar Pradesh worked on ring seine boats. With no prior experience of the coast or of fishing, their need for income and employment had caused them to seek work as crew:

> I came to this area eight years ago with my uncle and found fishing lucrative. I moved here, as there is no work at home. We have rooms in a large house, sharing ten to a room. We cook [for] ourselves. We buy wheat from the neighbours, as we don't have access to subsidized grains provided by the PDS [public distribution system] here. Our wives are involved in *bidi* [leaf-wrapped cigarette] rolling to earn some extra money at home.
>
> (Bhaiya, male, age 40)

The migrant workers did not have security in terms of their earnings, their living conditions were poor, and although they were polite about the boat owners, they admitted to experiencing stigma as *Hindikara* (Hindi-speaking people). No agents are involved in the migrant workers' recruitment. Rather, the migrants follow their kin and relatives in search of work: 'The season is only for six months. We then go home to help with farming. When we return, other boys come with us' (Bhaiya). While earning some income, the migrant workers are not integrated into the caste-based occupation on equal terms, which makes their labour precarious and insecure (Rao and Mitra, 2013; Rigg et al., 2016; Standing, 2011).

A further dimension of the labour mobilization was generational. In a focus group discussion with young men, two participants were candid about their feelings of intergenerational injustice and lack of recognition:

> We did not like to work with our previous unit, as we youngsters were dissatisfied with the behaviour of some of the elders in the unit. They are our relatives, yet they tried to extract work from us and kept abusing us for not working hard enough.
>
> We did not like it. Some of us have worked on ring seine boats for almost five years, yet they constantly undermined our experience and skills, so we decided to start our own new ring seine unit.

A key informant confirmed that the young men were operating their boats successfully and in Periyapattinam such boats were popularly called boys' nets (*pasanga valai*). While earlier they were ridiculed, they are now recognized as capable male fishers.

## The meanings of *sharing* in labour relations

Krishnamurthi, male, age 40, from Uppanankuppam, described a process of capital and labour mobilization similar to that described by Raman. He elaborated on the sharing of income and expenditure as follows:

> After the ban period this year, we launched the boat. We went for *thangal* fishing [stay fishing]. The expenditure was high[5] but the catch was good on our first trip – tuna and skipjack worth 600,000 rupees [USD8446], and other fish worth 200,000 rupees. We deducted the operational costs from the net amount, after the payment of the auctioneer's commission, and shared the income on a 60:40 basis, 60% for owners and 40% for the workers. The driver gets an extra half share. In the following trips, we made huge losses, and were unable to even recover our costs. We have an outstanding debt of Rs6 lakhs [INR600,000, USD8446].

While the general discourse was one of equal sharing of profits between the owners and workers, this was clearly not the case in practice, as the owners received two shares, one as owners and one as workers. There are additional ways of undercutting the workers' share:

> About 60% to 70% of the owners are genuine in sharing the income along these lines; others cheat the labourers by inflating the operational costs. For example, even if all of the diesel is not used, it is charged to the trip, liquor is added, and so on. The labourers from northern India

> work for even lower shares than the local labourers because they are new and don't have the skills.
>
> (Manikkam, male, age 53)

Earlier, few workers were involved in small-scale fishing. Rather, the boats were operated by kin, such as two brothers or a father and son. Today, especially on steel boats, typically there are five owners and 35 labourers, which has given rise to visible class relations. Such class relations were muted in the past, if they existed. The northern Indian labourers clearly have the worst deal, as they do not have any backup support in terms of household assets in the two villages. In 2017, Krishnamurthi explained that the ring seine season collapsed after a first bumper catch, which led to huge financial losses. The migrant workers were unable to withstand the losses, and many returned to their villages in the state of Uttar Pradesh, despite feeling that they had failed their families in providing income. After their return home, some worked on their small plots of land, while others survived by taking on temporary work, pulling rickshaws.

Because each boat has more than one owner, the dominant capitalist status is more broadly shared. However, as Raman pointed out, there is a desire to restrict the size of the shareholder group, such that each shareholder can earn higher incomes. The sense of precarity due to potential exclusion can lead to conflicts in the group, as indicated by Murthi, male, age 55, from Uppanankuppam:

> There is often a perception that the accountant has manipulated the accounts in favour of his own family, which creates splits in the group. Once the view that a small coterie is looting the rest spreads, it is hard to continue, as labourers too are unwilling to work on the boat.

The main issue appears to be growing competition and the need to prioritize individual interests over collective interests. Apart from a class divide, wherein the labour of migrant crew is appropriated for capital accumulation by the boat owners, who draw on caste and ethnic identities, we also found generational and gender divides. Most of the elderly male fishers were critical of the use of ring seines as destructive fishing gear and of the younger generations of fishers who were rapidly opting to use ring seines. Regardless of the method, the elderly fishers were of the opinion that the fish caught should be shared more widely rather than being concentrated in the hands of a few, because it would protect the livelihoods of fishing families that fish on small scale and thus maintain a sense of ethics and equity, which in the past was regulated by community institutions. In the case of women, the new technologies marginalized them from the fishing enterprise, and the only way for them to retain a share in the enterprise was by contributing capital, which sometimes had to be borrowed.

## Class, generation, and gender: some concluding thoughts

In this chapter, we have aimed to demonstrate how the capitalization of a traditional sector leads simultaneously to new opportunities for some and growing precarities for others and how small-scale fishers respond to the changes. In India, differentiation existed in marine fisheries before the capitalization of the fishing industry. The mechanized trawling sector, encouraged by state initiatives since the 1960s, was already in control of a disproportionate share of the fish catch. However, the majority of fishers (70%) remained in the small-scale sector, where they worked in accordance with principles of social equity and ecological sustainability, similar to those in the agrarian sector, which is dominated by small and marginal farmers (Sathiadhas, 2009).

With growing perceptions of injustice (Bavinck et al., 2018), along with the need for more resources for household reproduction and future aspirations, small-scale fishers have sought to challenge elite capital through group formation (especially by pooling financial and material resources). Expanding the base of boat owners and hence diffusing the concentration of capital locally has involved an increase in migration and mobility as strategies for accumulation (Sneddon, 2007). However, the outcomes vary for differently positioned people, with criteria for inclusion and exclusion drawing additionally on caste, ethnicity, generational, and gender identities.

The new class alliances are restricted to the fishing castes: in the case study villages of Periyapattinam and Uppanankuppam, the Pattinavars are unwilling to allow other castes to join as partners or shareholders. Greater capitalization has meant a greater demand for labour, which in the studied villages has been met by migrant workers, both from the agricultural hinterland and from more-distant and less-developed parts of northern India (Rao and Mitra, 2013). Despite the discourse of equal treatment, their labour is expropriated to build and sustain the capital of the new 'elite' shareholders, justified on the grounds of skill and experience. However, even among these new elites, there is growing competition resulting from declining fish catches and reducing incomes. Hence, they seek to establish smaller groups of partners, thus concentrating capital in the hands of a few and eroding the notion of sharing in the fishing community. This in turn, together with the adoption of increasingly risky and less sustainable strategies, gives rise to conflicts.

A relatively new form of differentiation and conflict is generational in character. Young men, in many cases well educated, are confronted with a lack of remunerative employment in either the public sector or the private sector. They aspire to return to fishing as boat owners, seek higher incomes than their fathers, and are keen to capitalize in order to earn comparatively more income in a shorter time frame. This leads to a willingness to overexploit marine resources through the use of destructive fishing gear, thereby contributing to long-term precarity, but also to the acceptance and reproduction of unequal gender and wider social relationships.

While some of the changes have opened up new opportunities for women, such as in auctioning, the small-scale trading activities of the majority of women are increasingly threatened by their inability to access fish. Both capitalization and the recruitment of male migrant labourers for subsidiary tasks have marginalized women's active role in fisheries, and dowry has emerged as an important source of both capital and status, and a means to compensate for their marginalization. Women are increasingly pawning their gold jewellery (as also mentioned in Chapter 6, this volume) to support this process of capital accumulation, which leaves them without any assets for coping in any future crises. The process of capital accumulation thus appears to conflict with processes of gender equality.

In the absence of adequate financing from banks and other public financial institutions, men who are unwilling to demand large dowries, raise resources for initial capital investment in boats or shares in boats by undertaking short-term migration for work overseas, working for relatively low wages in poor conditions (Gidwani and Sivaramakrishnan, 2003), or having to incur debt, mainly from private moneylenders at high rates of interest. Migration and labour mobility are crucial in this process of accumulation but have varied implications on enhanced security or precarity (Breman, 2007) for different groups of people, shaped by their social identity and position in caste and class hierarchies.

In summary, we found that unequal power relations and inequalities in the study villages were both challenged and reproduced through formal and informal institutions governing property rights, access to finance and capital, the availability of remunerative employment, and decent work. While there has been some diffusion of capital in the fishing caste, potentially contributing to a degree of security, it remains exclusive to that group, whereas outsiders, especially those in the labouring classes, face growing exclusion and precarity.

## Notes

1 Exports were facilitated through the establishment of seafood processing units with modern machinery for freezing, and this added value has contributed to an exponential growth in seafood exports from India of more than 55 times between 1961 and 2012, from 15,762 metric tonnes to 862,021 metric tonnes (MPEDA, 2018).
2 In line with ethical procedures to ensure confidentiality, the names of the villages and study participants have been changed to pseudonyms.
3 All quotes from the study participants were translated from Tamil by the authors of this chapter.
4 Either Caterpillar brand or Chinese brands of engines with a capacity of 500 HP.
5 According to Raman, the costs were 250,000 rupees [USD3519] for 4000 litres of diesel; 22,000 rupees [USD310] for 200 pieces of ice; 30,000 rupees [USD422] for rations, which included 15,000 rupees [USD211] for liquor; 2500 rupees [USD35] for water; and 1000 rupees [USD14] for other expenses.

## References

Appadurai, A. (1986). *The social life of things: Commodities in cultural perspective*. Cambridge: Cambridge University Press.

Bavinck, M. (2001). *Marine resource management: Conflict and resolution in the fisheries of the Coromandel Coast*. New Delhi: Sage.

Bavinck, M. (2008). Collective strategies and windfall catches: Fisher responses to tsunami relief efforts in South India. *Transforming Cultures*, *3*, 76–92. Retrieved from https://epress.lib.uts.edu.au/journals/index.php/TfC/article/view/923 accessed 19/12/2019.

Bavinck, M. (2014). Investigating poverty through the lens of riches – Immigration and segregation in Indian capture fisheries. *Development Policy Review*, *32*(1), 33–52.

Bavinck, M., Jentoft, S., and Schoeltens, J. (2018). Fisheries as social struggle: A reinvigorated social science research agenda. *Marine Policy*, *94*, 46–52.

Bavinck, M., and Vivekanandan, V. (2017). Qualities of self-governance and wellbeing in the fishing communities of northern Tamil Nadu, India – the role of Pattinavar ur panchayats. *Maritime Studies*, *16*, 16. Retrieved from https://link.springer.com/article/10.1186/s40152-017-0070-8 accessed 19/12/2019.

Breman, J. (2007). *The poverty regime in village India: Half a century of work and life at the bottom of the rural economy in south Gujarat*. Oxford: Oxford University Press.

Campling, L., Havice, E., and McCall, P.H. (2012). The political economy and ecology of capture fisheries: Market dynamics, resource access and relations of exploitation and resistance. *Journal of Agrarian Change*, *12*(2–3), 177–203.

Cresswell, T., and Merriman, P. (2011). *Geographies of mobilities: Practices, spaces and subjects*. London: Ashgate.

Gidwani, V., and Sivaramakrishnan, K. (2003). Circular migration and the spaces of cultural assertion. *Annals of the Association of American Geographers*, *93*(1), 186–213.

Government of India. (2017). *National policy on marine fisheries 2017*. Retrieved from http://vikaspedia.in/agriculture/policies-and-schemes/fisheries-related/national-policy-on-marine-fisheries-2017 accessed 19/12/2019.

Government of Tamil Nadu. (2017). *Tamil Nadu human development report 2017*. Retrieved from www.spc.tn.gov.in/tnhdr2017.html accessed 19/12/2019.

Kandiyoti, D. (1998). Gender, power and contestation: Rethinking bargaining with patriarchy. In C. Jackson and R. Pearson (Eds.), *Feminist visions of development: Gender analysis and Policy* (pp. 135–152). London: Routledge.

Lawrence, T.N., and Bhalla, R.S. (2018). Spatially explicit action research for coastal fisheries management. *PLoS One*, *13*(7), e0199841. doi:10.1371/journal.pone.0199841

MPEDA. (2018). *Marine products exports*. Marine Products Export Development Authority, India. Retrieved from www.mpeda.gov.in/MPEDA/marine_products_exports.php# accessed 19/12/2019.

Pillai, N.G.K., and Katiha, P. (2004). *Evolution of fisheries and aquaculture in India*. Retrieved from http://eprints.cmfri.org.in/23/ accessed 19/12/2019.

Rajan, J.B. (2002). *Labour mobility in the small-scale fisheries sector of Kerala*. Discussion Paper No. 44. Thiruvananthapuram: Centre for Development Studies.

Ram, K. (1991). *Mukkuvar women: Gender, hegemony and capitalist transformation in a south Indian fishing community*. London: Zed Books.

Rao, N. (2013). Migration, mobility and changing power relations: Aspirations and praxis of Bangladeshi migrants. *Gender, Place and Culture*, *21*(7), 872–887.

Rao, N., and Mitra, A. (2013). Migration, representations and social relations: Experiences of Jharkhand labour to western Uttar Pradesh. *Journal of Development Studies*, *49*(6), 846–860.

Rigg, J., Oven, K.J., Basyal, G.P., and Lamichhane, R. (2016). Between a rock and a hard place: Vulnerability and precarity in rural Nepal. *Geoforum*, *76*, 63–874.

Sathiadhas, R. (2009). Inter-sectoral disparity and marginalization in marine fisheries in India. *Asian Fisheries Science*, *22*, 773–8786.

Sathianandan, T.V. (2017). *Marine fish production in India – Present status*. Retrieved from http://eprints.cmfri.org.in/12164/1/2-Marine%20fish%20production%20in%20India.pdf accessed 19/12/2019.

Sen, A.K. (1990). Gender and cooperative conflicts. In I. Tinker (Ed.), *Persistent inequalities: Women and world development* (pp. 123–149). New York. Oxford University Press.

Sneddon, C. (2007). Nature's materiality and the circuitous paths of accumulation: Dispossession of freshwater fisheries in Cambodia. *Antipode*, *39*(1), 167–193.

Standing, G. (2011). *The precariat: The new dangerous class*. London: Bloomsbury.

Sundar, A. (2010). *Capitalist transformation and the evolution of civil society in a South Indian fishery*. Unpublished doctoral thesis, University of Toronto, Toronto.

Swaminathan, H., Suchitra, J.Y., and Lahoti, R. (2011). *Karnataka household asset survey: Measuring the gender asset gap*. Bangalore: Indian Institute of Management.

Vivekanandan, E. (2011). *Climate change and Indian marine fisheries*. Marine Fisheries Policy Brief No. 3. CMFRI Special Publication No. 105. Kochi: Central Marine Fisheries Research Institute.

Vivekanandan, V., and Kasim, H.M. (2011). *Fisheries management options for Tamil Nadu & Puducherry*. Chennai: Food and Agriculture Organization of the United Nations, World Bank, Government of Tamil Nadu, and Government of Puducherry.

# 3 Adapting to diminishing fish resources in Cambodia

## Fisheries on the shoulders of women and migrating adult children in fishing communities

*Kyoko Kusakabe and Prak Sereyvath*

### Introduction

Fishers are often considered the poorest of the poor. Fishing as a livelihood is seen as a last resort for the poor, and their marginalization makes it difficult for them to cope with reductions in fish catches (Baran and Myschowoda, 2009). However, as Béné and Friend (2011) and Coulthard (2012) note, fishers are heterogeneous. In the case of Cambodia, many people started to fish after the Democratic Kampuchea period (1975–1979) came to an end, since fishing provided them with a stable and sufficient income. With the recent decline in fishing resources (Johnstone et al., 2013; Teh et al., 2019), fishing has become a more precarious occupation. However, it is still the primary form of livelihood for fishers, who supplement their incomes in various ways, such as by borrowing money, migrating, and engaging in non-fishing and non-farming occupations (Mousset et al., 2016). In this chapter, we explore the coping strategies of fishing communities in Cambodia in the face of diminishing fish resources by focusing on how migration and activities that generate cash incomes work as adaptation strategies in three ecological settings: Tonle Sap lake, a floodplain, and a coastal area. By examining how different fishing communities have adopted different coping strategies, we are attempting to understand the fishers' agency: their preferential choices with respect to how they would like to construct their livelihoods.

The study was conducted in the period 2016–2018, in 14 community fisheries (CFs)[1] in Cambodia and comprised 191 interviews with members of fishing households, of whom 126 were women, and three focus group discussions in each CF. In this chapter, we examine one CF from each of the aforementioned three ecological settings (three CFs in total). We start with a description of the different fishing communities in Cambodia and then describe the conceptual framework used for the analysis of changes in the fishing households' livelihoods. Thereafter, we present the results of the analyses, including how women and their adult children are mobilized to support fishing livelihoods, and then compare and discuss the data from CF cases in each ecological setting.

## Fishing communities in Cambodia

Fishing provides full-time and part-time employment for up to 6 million people in Cambodia (Ministry of Agriculture, Forestry and Fisheries, 2011). The number of full-time fishers in Cambodia is quite low: only 2.5% of the population in 2005 (Lieng et al., 2018). Baran et al. (2014) note that in the 2008 national census, only 0.6% of population gave fishing as their primary occupation, 64% of all rural households were engaged in fishing, and the coastal areas had the highest dependency on fishing (for an analysis of data from the Cambodia Socio-Economic Survey 2011, see Chapter 12, this volume). Although the majority of those who live in rural Cambodia do not consider themselves fishers, most people do fish (Béné and Friend, 2011).

Tonle Sap lake covers an area of 3000 $km^2$ during dry season and 15,000 $km^2$ during the rainy season (Heng Sovannara, 2013). The lake's floodplain extends to 40,000 $km^2$. The coastline of Cambodia is 435 km (Heng Sovannara, 2013). The fisheries sector (harvesting, processing, and trade) contributes 8%–12% to the GDP, which is more than livestock (3.3%) (Ministry of Agriculture, Forestry and Fisheries, 2011). The total production of marine and freshwater fish in capture fisheries (excluding fish from rice field fisheries) was approximately 121,025 metric tonnes and 369,095 metric tonnes respectively in 2017.[2] Fisheries' production is estimated to be worth around USD200–300 million per year, and fish exports are valued at USD100 million per year (Ministry of Agriculture, Forestry and Fisheries, 2011). A study by Mousset et al., conducted in 37 villages in Cambodia, showed that the villagers consumed 3.6 kg of fish per week per household, which was more than double the amount of meat consumed (Mousset et al., 2016). In Cambodia, fish provide more than 60% of daily animal protein and approximately 20% of total protein consumption (Needham and Funge-Smith, 2015). Cambodia has one of the highest per capita consumptions of fish in the world: 63.15 kg per year (Needham and Funge-Smith, 2015). On average, fishing households earn USD333 per year from fishing, which provides a net income of USD170 (Mousset et al., 2016).

Before 2012, the government in Cambodia had demarcated the fishing grounds and designated fishing lots in order to give concessions to large operators. There were many conflicts between the owners of the fishing lots and fishers who operated on a smaller scale, especially after the 1990s, when more people started to fish (for more details about fishing conflicts, see Chapter 10, this volume). To establish fishing rights for small-scale fishers, the subdecree on CFs was issued in 2005 (Kingdom of Cambodia, 2005). Kurien hails the achievement of the establishment of the CFs in Cambodia as follows: 'In 2000, a bold initiative in Cambodia became a trail blazer when individual rights were replaced with community rights' (Kurien, 2018, p. 10). Although CFs are designated fishing areas for management, they do not have any budget or power to arrest illegal fishers. Under the 2005 subdecree, the regulations allow the use of only family-scale fishing gear but do not restrict the amount of fishing gear, and it has therefore been difficult for CFs to prevent overfishing (Chap et al., 2016; Sam et al., 2016) (for more information on CFs, see Chapter 10, this volume).

The official fisheries data do not show decreases in fish catches in inland and marine fisheries (Figure 3.1), but fishers have reported such decreases, especially since approximately 2012 (IUCN, 2016; Johnstone et al., 2013). Schwartz et al. (2016) state that 82% of their participants reported decreases in fish catches in the period 2000–2010. Although fish prices have increased, the prices of fuel and fishing equipment have risen too. Furthermore, An et al. (2008, cited in Schwartz et al. 2016, p. 2) found that the rate of increase in the price of fish was lower than the rate of increase in the price of rice. Various reasons have been suggested for the decrease in fish catches, including climate change, the use of illegal fishing gear, and increases in the numbers of fishers. Additionally, the abolishment of fishing lots in 2012 has led to a 'tragedy of the commons' and weak governance of the fishing resources (IUCN, 2016). Following dam construction in upstream areas in Cambodia, fishing incomes are expected to be further affected (Baran and Myschowoda, 2009).

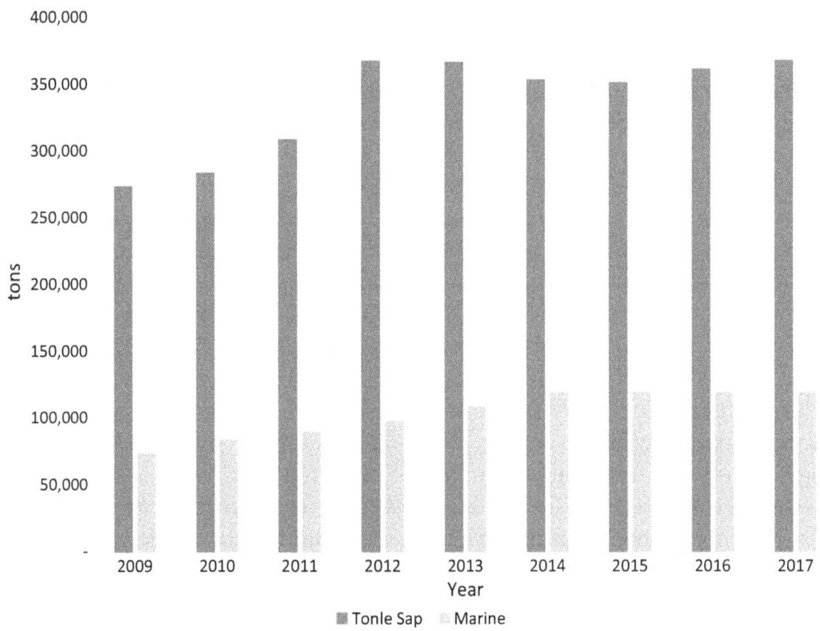


*Figure 3.1* Fish catches by small-scale and commercial fishers in Tonle Sap lake and in coastal areas, Cambodia

Source: Compiled from Ministry of Agriculture, Forestry and Fisheries' annual reports 2009 to 2018

With the decrease in catches and hence reduced incomes from fishing, fishing households are forced to seek alternative livelihoods. On the basis of data from the Cambodia Socio-Economic Survey 2011, Gunatilaka (Chapter 12, this volume) identifies households involved in fisheries as falling to within the poorest segments of the population. In a separate study, Schwartz et al. (2016) found that households with fishing as their primary occupation did not adapt successfully to environmental change, compared with other households. However, according to Béné and Friend (2011), there is no firm relationship between being poor and being a fisher. In a study of Tonle Sap, Teh et al. (2019) found slightly higher incomes for full-time fishers compared with the average incomes for rural populations. However, although the fishers might not be the poorest of the poor, they are still a relatively vulnerable group in Cambodia because they are susceptible to environmental change and have weak claims to resources (Teh et al., 2019). Béné and Friend (2011) argue that fisheries do not make people poor, but rather, the marginalization of fishing communities affects their well-being.

## Is migration for work and non-fishing work a diversification strategy for fishers?

In this section, we explore how fishing households cope in the face of a decline in fishing resources and whether and how migration is used as one of their adaptation strategies. According to the climate change literature, migration is an important adaptation strategy in the face of environmental changes. Some scholars note the positive effect of migration as an adaptation strategy (Islam et al., 2014; Saeed et al., 2016). New resources gained through remittances can contribute to the resilience of a livelihood in the place of origin (Gemenne and Blocher, 2017). Other scholars question the effectiveness of migration as an adaptation strategy, pointing to the economic and social costs of migration to both the migrants and the people left behind in the place of origin (Radel et al., 2018; Scheffran et al., 2011).

The effect of migration can be partly positive and partly negative. Radel et al. (2018) note that migration should not be seen as either a success or failure to adapt but rather as a reflection of the weak position of smallholders in the economy and in society. In the face of climate change, smallholders migrate, but they are often not able to adequately sustain their livelihoods (Gemenne and Blocher, 2017). This leads to the reinforcement of social inequality, since incomes earned by those who migrate for work only barely maintain the semi-subsistence farming system in Cambodia and elsewhere in the world. A similar finding was made by Morand et al. after their study of fishers in Mali, where migration as a strategy was not enough to reduce the fishers' vulnerability (Morand et al., 2012). Black et al. (2011) show that people may become more vulnerable as a result of migrating.

Bettini and Gioli (2016) point out that the new economics of labour migration (NELM) and sustainable livelihood analysis (SLA) frame migration as adaptation, thus emphasizing that households can improve their vulnerable situation through remittances without analysing why they were vulnerable in the first place. They also point out that there has been a lack of analyses of power structures that place smallholders at a disadvantage (Bettini and Gioli, 2016). As Radel et al. (2018) point out in the case of smallholders in Nicaragua, the NELM and forced migration literature often portray migration as a result of push factors such as poverty and low education, yet such a view can be apolitical and ahistorical because it does not take into consideration the weak power of such smallholders and the structural inequalities that limit their options in their place of origin. Felli and Castree (2012) note that the effect of the Royal Government of Cambodia's neoliberal policies that led to privatization, liberalization, and market-lead agrarian reform has been absent from the debate on migration as adaptation. Bettini and Gioli (2016) maintain the need for structural inequalities when discussing migration as adaptations. Migration is shaped not only by the economic needs of the migrants but also by other factors, including the state policy that regulates mobility (Massey et al., 1993), social and cultural networks (Castles and Miller, 1998), and other structural constraints (Hummel, 2016).

Viewing migration from a perspective of power structures is important in order to understand the gender differences in experiences of migration as an adaptation strategy. Women and men are affected differently by migration patterns (Aguilar, 2006; Jolly and Reeves, 2005). Tiwari and Joshi (2016) note the positive effect of men's out-migration: although women's workloads increased, both their access to education and their decision-making power increased too. However, to understand migration decisions as adaptation strategies from a gender perspective, it is important to analyse them with respect to how women and men define themselves and their roles. Wernersson (2018) argues that how farmers identify themselves defines the process of change in their adaptation practices. Mortreux and Barnett (2009) point out that it is important to consider how farmers perceive the costs and benefits of migrating and staying, since they involve more than simple economic calculations. People's decisions to use incoming resources depend on their perceived priorities. Guerin (2014) notes that debt is embedded in various feelings and emotions. Additionally, gendered obligations and ethical bonds influence how resources are used (Petchesky and Alexander, 2014). Furthermore, as Guerin (2014) notes, it is important to understand people's aspirations.

Therefore, if we are to understand fishers' adaptation strategies in the face of diminishing fish resources, we need to examine how households decide to use their incoming resources according to their perceived priority. When discussing how households adapt to shocks, it is important to explore the extent to which the women and men are trying to restore their livelihoods. For fishers, it is often the case that continuing fishing is the state in which they want to remain (Coulthard, 2012). Even among farmer-fishers, fishing is considered an indispensable source of supplementary income that is not abandoned, even when resources decrease (Pollnac et al., 2001). Pollnac et al. (2012) found that fishers with comparatively large households did not intend to leave their livelihood, and the analysis of Pollnac et al. revealed that this might have been because the fishers did not want to take risks. Xi et al. (2017) argue that households tend to maintain the same livelihood strategy if they are not forced by external factors to make changes. Their argument is based on their study conducted in rural Cambodia, but their study did not include fishing. Fishing households' decisions to continue to fish is closely linked to their identity, which further forms the basis of their social networks and sense of belonging (WorldFish Center, 2010). The WorldFish Center (2010) questions whether such identity constrains or supports women in fishing households, and it argues that women may subsidize men's fishing efforts with little return for themselves. In this chapter, we explore how women and their adult children are mobilized to support fishing occupations to uphold their identity as fishers, and we assess whether such decisions are based on productivity and returns rather than on perceived priorities.

## Fishing community on Tonle Sap

The fishing community in Cha village[3] lives on Tonle Sap lake. There are 220 households in the floating village, and people live on boats on the lake, moving their houses according to the water level in the lake. All of the fishers are full-time fishers. Although they do not have any land, they practise small-scale farming on the floodplain when the water levels subside. The number of fishers in the floating village started to grow when fish resources were abundant and provided a good source of income. They reported that they used to be able to catch 50–100 kg of fish per day in 2012, but by 2017, their catch dropped to 15–25 kg per day. Although fish prices were low when they were able to catch a lot of fish, the drop in fish catches was much larger than the rise in prices, and in 2017, they earned KHR10,000–54,000s (USD2.5–13.5) per day compared with KHR20,000–200,000 (USD5–50) per day in 2012. In Cha village, both women and men fish, although the women tend to stay on the home boat more often than men do, especially when their children are young. The supervision of children is important in the village because of the need to ensure that they do not fall into the water and drown. Furthermore, because the village is floating, it is difficult for the villagers to travel to hospitals or for their children to attend schools on the land. They do not have access to land, and their employment opportunities are more or less limited to fishing.

Living on floating houses is not cheap, as the villagers need to repair them every year to prevent disintegration. They also need a fishing boat as well as fishing net and traps in order to continue fishing. Fishers need to buy new fishing equipment every year, and when they do not have enough money, they use credit from intermediaries and repay them in fish.

Although Cha village is heavily dependent on fishing and fish resources have decreased by more than half, the villagers are keen to stay there. Ms Sophea, age 44, had 10 children, and one of her daughters migrated to Siem Reap city to work, while the other children remained at home. The household shared two boats and gill nets:

> We are fishers. What can we do outside fishing? We cannot stop fishing. Even when fish decreases, we will continue to fish. . . . There is fish in this village, and this is my birth place. I would like to die in this village. . . . If we stay here, we can only do fishing. . . . If children do not want to go out to work in other places, they can stay at home and fish together.[4]

When fishers face diminishing fish resources, they increase their fishing efforts and decrease their consumption (Schwartz et al., 2016). This was the case for Ms Thida, age 31, who had three young children, and her sole source of income was from her husband's fishing. He had to fish for longer hours than previously in order to catch the same amount of fish. The study participants said that they also borrowed money and sought other income sources. However, in practice, it was not easy for them to borrow money, because they needed to have particular types of relationships with their creditors in order to borrow – a common point noted by Petchesky and Alexander (2014). Ms Sophea had an ID card that showed her status and that of her household as 'poor', and therefore, it was hard for her to borrow money, even from the fish collector[5] who regularly bought fish from her.

Fishers who want to borrow money enter into a subordinate power relation with the fish collectors. There are no written contracts for such debts, but the borrowers are under a heavy obligation to make their repayments. Ms Sophea said that when she was able to borrow money from the fish collector, she was scolded by them if repayments were not made regularly. Therefore, she had to endure the degrading experience of being reminded that she was in a lower position than the collector: 'I borrow money from the collector and so I have to sell fish to her. I cannot sell to other merchants. If we sell to other merchants, she will be very angry' (Ms Sophea).

Despite the unequal power relations of fish collectors and fishers, the fishers in Cha village wanted to maintain them. As Ms Chinda, age 56, explained, despite the inequality in the relationships, the fishers believed that maintaining their one-to-one relationships with their fish collectors would provide them with some guarantees when they wanted to sell their catches. She lived with her son, who was a fisher, and although she had two daughters who had migrated to Siem Reap for work, 70% of her income was still earned from fishing. Thus, maintaining a regular buyer was important to her: 'It is not good to sell collectively, since each person has her own regular buyer. If we had to have a fixed buyer, we would not be able to sell at a better price' (Ms Chinda).

Apart from obtaining loans from intermediaries, the youths in Cha village migrated to work as labourers (e.g. in factories) in large cities in Cambodia, Thailand, and Korea or on farms in Thailand and Korea. Cha has a relatively long history of villagers' migrating to other cities in Cambodia, including Siem Reap and Phnom Penh. More recently, they have practised cross-border migration. Their remittances allow fishers to continue fishing. Migrant adult children return to live in the village from time to time, and some return to continue fishing. Although Ms Sophea had only one daughter who had migrated, the daughter's remittances had made a difference:

> Life was good when [the] fishing lot was there, but it became better when my daughter went to Siem Reap to work. I use remittances from my daughter to buy rice as well as nets and fishing equipment.

Parents depend on their adult daughters for remittances to support the education of their other siblings. Daughters are not considered core contributors to fishing households, and therefore, they are expected to contribute by earning away from home:

> The two daughters in Siem Reap will send remittances to us around 20 dollars [USD] per time, two or three times a year. I use this money for living expenses, repairing boats, and sending money to children who are studying, but I am not dependent on the remittances. Even if the daughters do not send any money, I will be able to make ends meet. Daughters do not fish, so even if they are here, it will not be any help for fishing. So, it is better to send them to study and let them work outside.
>
> (Ms Chinda)

By contrast, the roles of wives are limited to supporting their household's livelihoods on a daily basis, through helping their husbands when fishing and doing housework. Ms Mealea, age 31, had two young children and fished with her husband, using hooks and traps. Their fish catches decreased from 20 kg per day in 2012 to 10 kg per day in 2015. Ms Mealea was interested in migrating for work, but her husband did not agree with her:

> I really want to go to work in Thailand, but my husband does not agree and does not allow me to work in Thailand. He does not agree because there will be no one to look after the children.

Thus, in Cha village, fishing as a form of livelihood is sustained by women who borrow money and adult daughters who migrate for work and send remittances. Hence, livelihood resilience is both a gendered and generational effort in the village.

## Farming-fishing community on the floodplain

Russei village, in Banteay Meanchey Province, is situated on the Tonle Sap floodplain facing the border with Thailand. There are approximately 400 households in the village, of which 250 are fisher households. The main occupation of the villagers is rice farming. Although fishing is a secondary occupation, it is still crucial, because fish are one of the main sources of the villagers' nutrition. Almost all of the approximately 400 households have one or more members who engage in fishing in one form or another. The fishers travel to water bodies either by motorbike or bicycle, and the majority fish without a boat in ponds, rivers, and paddy fields. There are more than 100 natural ponds in Russei village. A few people have boats, which they use for fishing on ponds and rivers. As farming has never provided enough food for the farmers' families, farmers relied on fishing to make ends meet.

A canal was constructed in Russei village in 2015, and subsequently dry season rice production led to a reduction in water levels in the local reservoir, which in turn affected the fish habitats. Fish catches are decreasing, so cash incomes from fishing have declined. The participants noted that in 2007 they were able to earn approximately KHR126,000 (USD31.5) per day from fishing; in 2014, they earned KHR60,000 (USD15) per day; and in, 2017 they were able to catch only sufficient quantities just for home consumption or up to KHR25,000 (USD6.25) per day at maximum. Hence, as a source of supplementary income, fishing has declined. Ms Sarom, age 38, did not have any paddy land, and income from fishing used to be her household's main source of income. However, when the number of fish available started to decrease after c. 2010, she and her husband decided to work in Thailand for three years. They eventually returned to Russei, but since the fishing stocks were still poor, they stopped fishing and started to rent a paddy field:

> Before, it was possible to live only on fishing in this village. Even without migrating, one could live in this village only on fishing. Now, living only on fishing is not possible. By only fishing, we cannot repay loans and so we need to go for labour migration.
>
> (Ms Sarom)

In general, the villagers in Russei do not collect shells and snails in shallow waters, because it is considered a low status occupation and is looked down upon as an activity done only by the poorest people. When we asked questions about gleaning, the participants appeared slightly embarrassed and told us that they just did gleaning for home consumption and that it was not done for commercial purposes.

Given Russei's proximity to Thailand, many villagers migrated for work there, and the numbers of such migrants are increasing as incomes from fishing are decreasing: household members in approximately 90% of households migrated to Thailand. The number of migrants has increased since 2011. Income from remittances has become a crucial source for households' survival. The households' management of their finances depends on household members' migration for work. The villagers borrow money to build houses and repay the loans by using remittances, or they borrow money to pay for legal documents to enable them to work in Thailand. Ms Sreymom, age 59, had a husband who fished, and all of their five adult children had migrated to Thailand for work. Until 2010, she sold fish every day, but when the interview was held with her, there were no longer enough fish for her to sell:

> The causes of migration are poverty, less fish and [the hope that the migrants] might earn more money to support the family. My children send a remittance of 5000 baht [USD166 USD] per month for living expenses and for paying back the loan to bank.
>
> (Ms Sreymom)

The remittances are often used to invest in fishing to supplement incomes from other sources. Mr Vuthy, age 58, had four adult children, all of whom worked in Thailand. He used their remittances to buy more paddy land and a boat, a trap, and a net. In total, he had 5 ha of paddy land, and although he continued to fish, he did not catch enough to sell:

> Before, my income was from fish and farming. Both were almost the same. With remittances from my children, I bought extra land and a boat. With this, I am now able to increase income from farming and fishing.
>
> (Mr Vuthy)

Migration can be seen as a form of social debt. Adult children, especially daughters, are expected to earn and send remittances as a form of filial piety. In return, parents take care of their grandchildren. Mr Chap, age 58, had three sets of gill nets and 10 fish traps. He spent THB1000 (USD33) per year on fishing equipment every year. Additionally, Mr Chap bought a hand tractor in 2010 and a pumping machine and motorbike in 2017. All of the purchases were made with remittances from his six adult children who had migrated to Thailand in 2008. Of Mr Chap's total household income, approximately 60% was from remittances, while his income from fish and rice was only 10% in each case. In return for the children's remittances, he and his wife took care of their five grandchildren.

## Coastal fishing community

Thmei, in Kampot Province, is a coastal village with 747 households, 80% of which are occupied by fishers who catch fish, crabs, and shrimps. When the research was conducted, the average income from fishing per household was KHR50,000–100,000 (USD12.5–25) per day. Most villagers in Thmei cultivate rice and raise livestock, in addition to fishing. Only approximately 20 families do not have a rice field. There is gender discrimination in the wages paid to villager who work as daily wage workers in fishing: men are paid KHR40,000 (USD10) per day, whereas women are paid only KHR20,000 (USD5).

Thmei village has suffered from decreased incomes from fishing compared with those in 2011 and 2012. Those who fish on a small scale and do not have boats to rely on, use push nets (a net type used mainly in shallow water), and earn only KHR20,000 (USD5) per day, whereas they used to earn KHR55,000 (USD13.75) per day. Some fishers with boats used to earn KHR450,000 (USD112.5) per day in 2011 and 2012, but this has since decreased to KHR150,000 (USD37.5) per day. Some used to catch shrimps and earned KHR18 million (USD4500) per year selling dried shrimps but can earn only KHR3.45 million (USD862.5) per year.

Although incomes from fishing have decreased, they are still higher than earnings from working in a factory. The relatively good income from fishing encourages fishers to continue to fish and to supplement their income with money from other sources. Fishers do not want to abandon fishing, because it is a quick way to earn money. Ms Phany, age 57, had a husband and three adult sons who fished and one son and a daughter who had migrated to work in Phnom Penh. Her husband and sons used push nets, which were one of the least profitable ways of catching shrimps in Thmei. Nevertheless, they preferred to continue to fish:

> Although the fishery resources are declining, we still go fishing because if we do not go, there will be nothing to eat. If we depend on rice cultivation and remittances from children, it is not enough.
>
> (Ms Phany)

The decrease in fish resources occurred at the same time as the establishment of a garment factory and casino in the vicinity of the village. Young people work in the casino and in factories to support their household, but their incomes are still considered lower than income earned from fishing. Often, women work in the factories and men fish, and the men have higher incomes. When working as fish labourers, men earn USD371 per month, whereas women in garment factories earn USD170 per month (youths' FGD). Factory work is not considered an 'occupation' in the sense of established work (*kangie chabah roah*), and it has a lower status than fishing has:

> We can earn 300 USD per month from factory jobs, or a minimum of 150 dollars [USD]. More women work in the factory than men . . . but the income at present is low and we are paid only once a month for this work, while fishing can give a daily income.
>
> (Participant in women's FGD)

In the coastal area, the gender gap in earnings is large, although it has improved a little. Earlier, widows had few options to make a living. For instance, Ms Nary, age 32, said that the only way her mother could earn an income after becoming a widow was by working in a salt field and by collecting krill. Working in the salt fields is considered degrading and is not done by the local people, because the work is physically hard and low paid. Ms Nary said, 'Nowadays, women can work in factories, hotels, restaurants, and migrate'. Women's livelihood choices have increased, and their role in supporting their family has also increased:

> Nowadays, women go out to earn income. Before, women did not go, but nowadays, they think that if both husband and wife work, then they can get double income. So, both of them go together. In construction work, women's wages can be lower than that of men's, but in some work, women can earn more than men. Nowadays, men rest during Saturdays and Sundays, but women work every day. Before, women used to be at home, but now more women than men are going out to earn money.
>
> (Participant in women's FGD)

Thus, women are taking on more work to support the fisher households' livelihoods. This can be seen as women either converting their time from household work (reproduction) to external work (production) or adjusting their time to create assets to support the fishing occupation.

## Concluding discussion

When faced with a decrease in fishing resources, households adapt in different ways. In Cha village, fishers borrow money and migrate. The marginalization of Cha in terms of access to services (health and education and sources of employment) has led fishers to adopt a coping strategy that subordinates them to fish collectors, to whom fishers are indebted.

In Russei, the villagers' coping strategy is mainly migration. This is also the case in Cha, but the strategy is used to a lesser extent than it is in Russei. In Russei, remittances have replaced income from fish, and some villagers have stopped fishing altogether. Due to resource depletion, adult children play a large role in supporting the villagers' livelihoods. Daughters are expected to bear such responsibilities to a greater extent than are sons.

By contrast, in Thmei village, there are many options for earning cash, including work in nearby factories and the casino, while fishing remains the most lucrative income-generating activity. However, the decrease in fishing resources has encouraged fishers to look for additional sources of income. Women, who previously stayed at home, are employed in greater numbers than men in factories and casinos. However, their income is lower than that of male fishers and is considered a minor contribution to the household.

As noted by scholars such as Wernersson (2018) and Coulthard (2012), fishers' identities are shaping adaptation decisions, and women and their adult children are being mobilized to earn cash to support fishing as a livelihood. The decisions convert women's time resources into financial resources to support fishing so that they can earn a small amount of income through hired labour. Gender norms in fishing households mean that women are treated as though they have an elasticity of time because they ensure that women contribute to financial investments in fishing as a livelihood while they also have to juggle their multiple roles in the households. Adult children, especially daughters, are expected to convert both their time and their filial piety into financial resources to support fishing as a livelihood. The WorldFish Center (2010) questions whether women's role is supportive, complementary, or subsidizing. In this chapter, we have shown that women 'subsidize' fishing efforts through their various incomes. Women and their adult children earn cash and invest it in fishing, which does not directly benefit them or help to ease their workload. Instead, women work harder to maintain the fishing profession of their household.

Since the purpose of women's income is to support the existing livelihoods in households, their income does not necessarily lead to their recognition, their empowerment, or any degree of freedom to expand their economic activities. This explains why, despite the changing livelihoods, the gendered division of labour and gender relations remain firmly intact. Notably, in all three ecological settings, although women's roles are changing and women play a large role in supporting fishers' livelihoods, there seems to have been little improvement in women's status in the households and in the community. Although the livelihood system is being maintained by women, they are still expected to continue their role in taking care of the house.

Our analysis of fishers' coping strategies in the face of diminishing resources through an investigation of how non-fish income flows are used and/or invested shows that other resources are mobilized by women to support the existing livelihood system. Since the mobilization of resources is meant to maintain the status quo, changes in roles and activities do not challenge existing gender relations, even when women play a large role in maintaining household finances. Furthermore, the low salaries earned by women who work in factories do not challenge existing gender norms. A household's identity as a fishing household is based on the occupation of the men in the household. Investment choices are made to prioritize their occupation, thus making the whole coping strategy gendered. Women continue to engage in lower-paid jobs as part of the system to support men's fishing activities. Migration for work by household members, especially adult daughters, is one such coping strategy, but it also typically reflects the household decisions that daughters should finance fishing activities and thus uphold the household's identity as fishers. In summary, this chapter has highlighted the importance of understanding fishers' identities and priorities in their investments in their livelihoods in order to better capture why women and men take up different types of work and how such priorities function to maintain the gender relations in the household in the face of changing fish resources and risks in the fishing occupation.

## Notes

1 According to Chapter 2 Article 6 of the subdecree on community fisheries (Kingdom of Cambodia, 2005),

> A community fisheries . . . is a group of physical persons holding Khmer citizenship who live in or near the fishing area, voluntarily established and taking the initiative to improve their own standard of living by using and processing fisheries resources sustainably to contribute to economic and social improvement and poverty alleviation.

Community fisheries need to be approved by the Fisheries Administration (in the Ministry of Agriculture, Forestry and Fisheries) in order to be recognized as a CF.

2 Ministry of Agriculture, Forestry and Fisheries 2018 report written in Khmer on fisheries statistics, accessed from the ministry in February 2019.

3 To preserve anonymity, the names of the villages and study participants are pseudonyms.

4 All interviews and FGDs were held in Khmer. Some of the translations in this chapter were done by Kyoko Kusakabe (the first author of this chapter) and some by Khmer – English translators (since the interviewer did not speak English).

5 Fish collectors buy fish directly from fishers and then sell the fish to fish wholesalers or other intermediaries.

## References

Aguilar, L. (2006). *Climate change and disaster mitigation: Gender makes the difference.* Retrieved from www.bridge.ids.ac.uk/global-resources/resource/A53612?lang=en accessed 19/12/2019.

An Channthla, et al. (2008). Big fish catch, but small gain. *Economics Today, 2*(26), November 11, 2008.

Baran, E., and Myschowoda, C. (2009). Dams and fisheries in the Mekong Basin. *Aquatic Ecosystem Health and Management, 12*(3), 227–234.

Baran, E., Phen, C., Vuthy, L., Nasielski, J., Saray, S., Bunthang, T., . . . Sokhom, T. (2014). Fish resources in Cambodia (2001–2011). In Save Cambodia's Wildlife (Ed.), *Atlas of Cambodia: Maps on socio-economic development and environment* (pp. 37–48). Phnom Penh: Save Cambodia's Wildlife.

Béné, C., and Friend, R. (2011). Poverty in small-scale fisheries: Old issue, new analysis. *Progress in Development Studies, 11*(2), 119–144.

Bettini, G., and Gioli, G. (2016). Waltz with development: Insights on the developmentalization of climate-induced migration. *Migration and Development, 5*(2), 171–189.

Black, R., Bennett, S.R.G., Thomas, S.M., and Beddington, J.R. (2011). Migration as adaptation. *Nature, 478*, 447–449.

Castles, S., and Miller, M.J. (1998). *The age of migration: International population movements in the modern world.* New York: Guilford Press.

Chap, S., Touch, P., and Diepart, J.C. (2016). *Fisheries reforms and right-based fisheries: Insights from community fisheries across Cambodia*. Phnom Penh: The Learning Institute.

Coulthard, S. (2012). Can we be both resilient and well, and what choices do people have? Incorporating agency into the resilience debate from a fisheries perspective. *Ecology and Society*, *17*(1), 4. Retrieved from www.ecologyandsociety.org/vol17/iss1/art4/ accessed 19/12/2019.

Felli, R., and Castree, N. (2012). Commentary. *Environment and Planning A*, *44*, 1–4.

Gemenne, F., and Blocher, J. (2017). How can migration serve adaptation to climate change? Challenges to fleshing out a policy ideal. *Geographical Journal*, *183*(4), 336–347.

Guerin, I. (2014). Juggling with debt, social ties, and values: The everyday use of microcredit in rural South India. *Current Anthropology*, *55*(9), S40–S50.

Heng Sovannara. (2013). *Overview of fisheries sector in Cambodia*. Retrieved from https://sustain.pata.org/wp-content/uploads/2014/09/HENG-Sovannara-Fisheries.pdf accessed 19/12/2019.

Hummel, D. (2016). Climate change, land degradation and migration in Mali and Senegal – Some policy implications. *Migration and Development*, *5*(2), 211–233.

Islam, M.M., Sallu, S., Hubacek, K., and Paavola, J. (2014). Migrating to tackle climate variability and change? Insights from coastal fishing communities in Bangladesh. *Climate Change*, *124*, 733–746.

IUCN. (2016). *Fisheries conservation and governance in the Tonle Sap*. Retrieved from www.iucn.org/news/cambodia/201611/fisheries-conservation-and-governance-tonle-sap accessed 19/12/2019.

Johnstone, G., Puskur, R., DeClerck, F., Mam, K., Il, O., Sithirith, M., . . . Rest, S. (2013). *Tonle Sap scoping report*. CGIAR research program on aquatic agricultural systems, Penang, Malaysia. Project Report AAS-2013-28. Retrieved from http://aquaticcommons.org/12823/1/AAS-2013-28.pdf accessed 15/1/2020.

Jolly, S., and Reeves, H. (2005). *Gender and migration: Overview report*. Institute of Development Studies (IDS). Retrieved from www.bridge.ids.ac.uk/sites/bridge.ids.ac.uk/files/reports/CEP-Mig-OR.pdf accessed 19/12/2019.

Kingdom of Cambodia. (2005). *Sub-decree on Community Fisheries Management*. Subdecree No. 80/OrNorKror/BorKor. Retrieved from www.ecolex.org/details/legislation/sub-decree-no-80ornorkrorborkor-on-community-fisheries-management-lex-faoc085864/ accessed 19/12/2019.

Kurien, J. (2018). From individual rights to community commons. *SAMUDRA Report*, *80*, 10–13. Retrieved from www.icsf.net/images/samudra/pdf/english/issue_80/270_Samudra_80_Dec_2018_ICSF.pdf accessed 19/12/2019.

Lieng, S., Yagi, N., Mori, A., and Hastings, J.D. (2018). Savings-group improvements contribute to sustainable community-fisheries management: A case study in Cambodia. *Sustainability*, *10*(8), 2905. Retrieved from https://doi.org/10.3390/su10082905 accessed 19/12/2019.

Massey, D.S., Arango, J., Hugo, G., Kouaouci, A., Pellegrino, A., and Taylor, J.E. (1993). Theories of migration: A review and appraisal. *Population and Development Review*, *19*(3), 431–466.

Ministry of Agriculture, Forestry and Fisheries. (2011). *The strategic planning framework for fisheries: 2010–2019*. Retrieved from www.cdc-crdb.gov.kh/cdc/

documents/Sector_Strategy/4_Fisheries/Fisheries_The_Strategic_Planning_Framework_2010_2019.pdf accessed 19/12/2019.
Morand, P.A., Kodio, A., Andrew, N., Sinaba, F., Lemoalle, J., and Béné, C. (2012). Vulnerablity and adaptation of African rural populations to hydro-climate change: Experience from fishing communities in the Inner Niger Delta (Mali). *Climate Change*, *115*, 463–483.
Mortreux, C., and Barnett, J. (2009). Climate change, migration and adaptation in Funafuti, Tuvalu. *Global Environmental Change*, *19*, 105–112.
Mousset, E., Rogers, V., Saray, S., Ouch, K., Srey, S., Mith, S., and Baran, E. (2016). *Roles and values of fish in rural welfare in Cambodia (welfare data analysis)*. Retrieved from http://pubs.iclarm.net/resource_centre/WF-4077.pdf accessed 19/12/2019.
Needham, S., and Funge-Smith, S. (2015). *The consumption of fish and fish products in the Asia-Pacific region based on household surveys*. RAP Publication 2015/12. Bangkok: Food and Agriculture Organization of the United Nations, Regional Office for Asia and the Pacific.
Petchesky, R., and Alexander, M. (2014). Introduction: Life and debt. *Women's Studies Quarterly*, *42*(1–2), 13–20.
Pollnac, R.B., Bavinck, M., and Monnereau, I. (2012). Job satisfaction in fisheries compared. *Social Indicators Research*, *109*, 119–133.
Pollnac, R.B., Pomeroy, R., and Harkes, I.H.T. (2001). Fishery policy and job satisfaction in three southeast Asian fisheries. *Ocean and Coastal Management*, *44*, 531–544.
Radel, C., Schmook, B., Carte, L., and Mardero, S. (2018). Toward a political ecology of migration: Land, labor migration and climate change in Northwestern Nicaragua. *World Development*, *108*, 263–273.
Saeed, F., Salik, K.M., and Ishfaq, S. (2016). *Climate induced rural-to-urban migration in Pakistan*. Working Paper. Retrieved from http://prise.odi.org/wp-content/uploads/2016/01/Low_Res-Climate-induced-rural-to-urban-migration-in-Pakistan.pdf accessed 19/12/2019.
Sam, S., Ky, C., Keum, K., Sarom, M., and Sok, R. (2016). *Common pool resources and climate change adaptation: Community-based natural resource management in Cambodia*. CDRI Working Paper Series No. 109. Phnom Penh: Cambodia Development Research Institute. Retrieved from https://cdri.org.kh/wp-content/uploads/wp109e.pdf accessed 19/12/2019.
Scheffran, J., Marmer, E., and Sow, P. (2011). Migration as a contribution to resilience and innovation in climate adaptation: Social networks and co-development in northwest Africa. *Applied Geography*, *33*(1), 119–127.
Schwartz, N.B., Gätke, P., and Baran, E. (2016). *Adaptation to environmental change among fishing-dependent households in Cambodia*. Phnom Penh: Inland Fisheries Research and Development Institute (Fisheries Administration) and WorldFish. Retrieved from http://pubs.iclarm.net/resource_centre/WF-4071.pdf accessed 19/12/2019.
Teh, L.S.L., Bond, N., Krishna, K.C., Fraser, E., Seng, R., and Sumaila, U.R. (2019). The economic impact of global change on fishing and non-fishing households in the Tonle Sap ecosystem, Pursat, Cambodia. *Fisheries Research*, *210*, 71–80.

Tiwari, P.C., and Joshi, B. (2016). Gender processes in rural out-migration and socio-economic development in the Himalaya. *Migration and Development*, *5*(2), 330–350.

Wernersson, J. (2018). Rethinking identity in adaptation research: Performativity and livestock keeping practices in the Kenyan drylands. *World Development*, *108*, 283–295.

WorldFish Center. (2010). *Gender and fisheries: Do women support, complement or subsidize men's small-scale fishing activities?* WorldFish Centre Issues Brief 2108. Retrieved from https://d3gxp3iknbs7bs.cloudfront.net/attachments/d345620e-a536-4f8f-b668-ec88dd0d3079.pdf accessed 19/12/2019.

Xi, J., Pouliot, M., and Walelign, S.Z. (2017). Livelihood strategies and dynamics in rural Cambodia. *World Development*, *97*, 266–278.

# 4 Seasonal migration, resource access, contestation, and conflict among fishers on the west and east coasts of Sri Lanka

*Nadine Vanniasinkam, Mohamed Faslan and Nireka Weeratunge*

## Introduction

Seasonal internal migration for fishing compels migrant and host communities to share access to the same fisheries resource, thus creating potential for conflict. Seasonal coast-to-coast migration is a widespread livelihood strategy on the west coast of Sri Lanka, induced by monsoonal weather patterns, seasonal fish movements between different parts of the island, and an implicit understanding both by the state and by fishing communities that 'the sea is for all'. Fishers on the west coast have been migrating to the northeast and east coasts of the island for periods of six to eight months annually for generations. Weather-related challenges to fishing in their home villages, local knowledge of fish movements, and the allure of rich fish resources off the northeast and east coasts have contributed to this process. As a consequence, most migrant fishers from the west coast occupy two homes: a permanent house in their own village and a temporary *wadi* or *wadiya* (camp) in the host village.

Before the beginning of civil war in 1983, migrant and host communities appeared to have shared amicable relations. However, during the war, from 1983 to 2009, communities became polarized. Moreover, horizontal hostilities over access to resources have been caused by the tsunami of 2004, coastal resource depletion, climate change, fisheries restrictions and regulations, postwar industrialization, and tourism development. These hostilities have resulted in contestations and disputes over rights to fish resources and sporadically escalated into violent conflict, with relational configurations shifting among migrant and host fishing communities that previously coexisted to a large extent. On occasions, the conflicts have manifested as protests, physical violence, and property destruction. The imposition of area-specific regulations and restrictions by fisheries authorities has added a vertical layer to the conflicts, challenging the customary practices of fishing communities, which are faced with few alternatives to earn a living.

In this chapter, we argue that contestation over resources is manifested in two discourses: migrant fishers' right to continue a tradition of migration and host fishers' right to their own resources. The aim is to unravel these perspectives to understand the contestation and disputes, which may erupt into conflicts between the two groups during coast-to-coast migration between two districts in Sri Lanka: Puttalam District and Trincomalee District. In doing so, we present a nuanced analysis of the individual actors and the socioeconomic, cultural, institutional, and political drivers of conflict among host and migrant fishing communities and between fishing communities and the state.

The chapter begins with an outline of the economic context of fisheries in the two study districts, followed by a brief discussion of the conceptual lens adopted and the methods used in data collection. This is followed by a historical overview of migration patterns and relations between migrant and host fishing communities. Thereafter, the core of the analysis of the dynamics of contestation and conflict among study communities on the west and east coasts is presented. The perceptions of depletion of and access to resources between migrant fishing communities and host fishing communities (actors who drive conflict) and how these resource-based contestations and disputes can manifest as ethnic and/or regional conflicts are discussed.

## Fisheries economy in Puttalam District and Trincomalee District in the national context

Despite Sri Lanka being an island, merely 1.3% of its GDP came from fisheries in 2016 (Ministry of Fisheries and Aquatic Resources Development, n.d.). This accounted for 7% of direct and indirect employment nationally. Marine fisheries contributed to 86.1% of Sri Lanka's fish catch, reaching 456,990 metric tonnes in 2016 (Ministry of Fisheries and Aquatic Resources Development, n.d.). Approximately 90% of Sri Lanka's fishing fleet comprises small-scale fishing vessels, which were responsible for approximately 60% of the marine catch in 2016 (Ministry of Fisheries and Aquatic Resources Development, n.d.).

Nationally, the marine fish catch more than doubled between 1983 and 2014, after plummeting in 2005 due to the tsunami in 2004,[1] and it has shown a slight decline since 2014 (Ministry of Fisheries and Aquatic Resources Development, n.d.). Puttalam has the largest fish catch of all districts in Sri Lanka, contributing 16.2% of the national fish catch, whereas Trincomalee District contributes 5.7% of the catch (Ministry of Fisheries and Aquatic Resources Development, n.d.). In the period 1983–2016, the marine fish catch rose by 2.9 times in Puttalam, compared with 1.8 times in Trincomalee (Ministry of Fisheries and Aquatic Resources Development, n.d.). There was a substantial decline in fish catches on the north and east coasts during the war, but the trend reversed after the war. Marine fish catches reached a peak in Trincomalee District in 2012 and in Puttalam District in 2015.

There has been a gradual shift away from traditional fishing vessels and beach seine fishing towards mechanized fishing vessels since the Blue Revolution in the 1950s (Amarasinghe, 1989). The introduction of new motorized fibre-reinforced plastic boats,[2] as well as new fishing gear and techniques, led to significant changes in economic and social relations. A slight decline in the fishing fleet has accompanied a decline in fish catches nationally, including in the study districts. By 2016, 48% of Sri Lanka's fishing fleet comprised small motorized boats and 35% comprised non-motorized traditional canoes and rafts[3] (Ministry of Fisheries and Aquatic Resources Development, n.d.). While both Puttalam (61%) and Trincomalee (56%) had higher proportions of motorized boats than that at the national level (Ministry of Fisheries and Aquatic Resources Development, n.d.), Puttalam had twice the number of small motorized boats and 1.7 times more traditional vessels than Trincomalee in 2016, despite a noteworthy increase in the Trincomalee fleet between 2014 and 2016.

Furthermore, the decrease both in the marine catch and in the size of the fleet is reflected in the decline in fishers and fishing households nationally by 2016. Before the decline, there had been a 2.1-fold increase in active marine fishers in Trincomalee District, compared with a 1.3-fold increase in Puttalam District between 2004 and 2015 (Ministry of Fisheries and Aquatic Resources Development, n.d.; National Aquatic Resources Research and Development Agency, n.d.). Thus, in 2016, Trincomalee District had 1.4 times as many active marine fishers and 1.6 times as many fishing households than Puttalam District did.

The fisheries statistics reveal that in 2016 Puttalam District, which had a fleet that was almost twice as large as the one in Trincomalee District and 30% fewer fishers, had three times the size of the catch in Trincomalee District (Ministry of Fisheries and Aquatic Resources Development, n.d.). However, the most recent stock assessment indicates a drastic reduction in fish stocks in the coastal waters all around the island over the past 30 years to one-fifth of the fish resources recorded in the period 1978–1980, with higher rates of depletion off the northwestern coast compared with that off the east coast.[4] Thus, inequalities between the two study districts in terms of catches, technology, and labour might partially explain the current contestation and disputes between migrant fishers who move from a more overfished fishery in Puttalam and host communities in which people make a living from a less overfished fishery in Trincomalee.

## Conceptual approach

Several scholars have challenged earlier monocausal theories of conflict over natural resources centred on scarcity, greed, or grievance, arguing that such conflicts are multicausal and multilevel and involve multiple actors (Bavinck et al., 2014; Frerks et al., 2014). In this broader and more nuanced approach, environmental factors are combined with sociopolitical factors to explain conflict.

A global literature exists on conflicts related to coastal and fish resources, including those associated with fisheries migrations (Adhuri, 2009; Salayo et al., 2006). While some scholars who have studied conflicts related to coastal and/or fish resources and migration have been concerned mainly with governance and the institutional factors underlying such conflicts, others have looked at both environmental and sociopolitical issues, including conflicts that manifest as ethnic (Abobi and Alhassan, 2015; Jul-Larsen, 1994; Siriwardena-de Zoysa, 2018).

In this chapter, we combine political ecology and social well-being approaches to look at the multiple factors that may contribute to conflict between migrant and host fishing groups in Sri Lanka. In a fisheries context, the political ecology lens used in this chapter is influenced by Fabinyi et al. (2015), who addressed local inequalities and perceived marginalities in fisheries governance in the Pacific, and by Bennett (2019), who emphasizes the role of power in the marine environment and the marginalization of small-scale fishing and indigenous coastal communities. Social well-being is defined as 'A state of being with others, where human needs are met, where one can act meaningfully to pursue one's goals, and where one enjoys a satisfactory quality of life' (McGregor, 2008, p. 1). As elaborated in Chapters 1 and 6 in this volume, social well-being has three dimensions: material (including environmental), relational, and subjective. The assumption in this chapter is that the underlying material, relational, and/or subjective motivations in the pursuit of well-being by members of fishing households in one community can conflict with those of another community, resulting in different well-being outcomes or ill-being along the three dimensions. The institutions that shape the rules and rights of resource access and use are part of the relational dimension of social well-being. The social well-being approach is used to understand how conflict over the fishing resource between migrant and host fishers is constructed through different well-being aspirations of the two groups and the outcomes for the two groups. Regulatory institutions can play a role in resolving or exacerbating tensions between these groups. Thus, the political ecology perspective helps us to understand the institutional inequalities in access to resources, whereas the social well-being approach provides an actor-centred approach to unravelling underlying motivations for conflict.

## Methods and scope

This chapter is based on the findings of a Sri Lankan study of four small-scale fishing communities, conducted by using qualitative and quantitative methods, as elaborated in Chapters 1 and 6.[5] The analysis presented here draws primarily on a literature review and qualitative data from livelihood portfolios; focus group discussions (FGDs) with men, women and youths; and in-depth interviews, including life history narratives, social network mapping for household members, and community and migration histories.

The study focused on the migrant villages of Cattiyur and Kadalpalli in Puttalam District and the host villages of Uppukadal and Selippur in Trincomalee District.[6] The villagers constitute three ethnic and religious groups: Sinhalese Catholics in Kadalpalli, Tamil Hindus in Cattiyur and Uppukadal, and Muslims in Selippur. Fishing locations, methods, and assets differ among the four villages, with a higher proportion of Kadalpalli and Selippur fishers owning motorized boats. Approximately 75% of fishers and family members of fishing households in the west coast study sites migrate seasonally.

## Historical overview of relations between migrant and host fishing communities

Seasonal migration, which is perceived as an adaptive livelihood strategy to maximize returns from marine resources around the coastal belt of Sri Lanka, has generated a culture of mobility among migrant fishing communities. Migration as a strategy can also be linked to kinship and caste, particularly among Tamil Hindus from the west coast who own beach seines, belong to the same Tamil Karaiyar caste, and migrate to the northeast and east coasts, where they also own beach seines. Another factor linking migrant and host communities is the Tamil language. Ethnographic evidence suggests a southern Indian origin for fisher castes on the west coast (Roberts, 1982). Fisherfolk from these communities have maintained a bilingual, hybrid identity, although they have increasingly identified as Sinhalese since the war. However, migrant fishers have been able to communicate and interact with their Tamil-speaking counterparts on the northeast and east coasts. Thus, relations between migrant fishers and host communities in Sri Lanka have been long-standing and were cordial before the war began in 1983, according to most study participants. Cooperation, rather than conflict, seems to have been the norm in relations between the two groups. Some migrants had intermarried with local people and established kin relations, while others had families, property, and ancestral graves on both coasts. Although historical data on the migration process is scant, records of migrants from the Negombo area on the west coast to the districts of Mullaitivu, Trincomalee, and Batticaloa on the northeast and east coasts exist in colonial administration reports from 1868 onwards.[7] The reports mention jealousy and occasional conflicts between migrant and local fishers, which were resolved by the relevant government agent in the Eastern Province. Migrants in the study locations were able to trace incidents of migration practised by their parents and/or grandparents.

The relations between the migrant fishers and host communities changed significantly due to the civil war (1983–2009), which resulted in a decline in migration from the west coast to the northeast and east coasts and in the weakening of ties with fishing communities on the east coast, particularly among the younger generations. Before the war, most migrants went to Mullaitivu District. Most of those from the west coast who continued to migrate during the war concentrated their activities in Trincomalee District, due to the presence of the navy. They lost the trust and goodwill of host fishing communities due to their ability to negotiate checkpoints and fishing restrictions with the army and navy, at a time when fishing communities in the northeast and east were suffering deprivations and fishing bans. Ethnic polarization widened when west coast fishers increasingly began to underplay their previously fluid and/or hybrid identities and to assert a Sinhalese ethnic identity, which facilitated relations with the armed forces. Furthermore, those migrants had little or no interaction with host communities, due to the displacement of the latter to refugee camps in neighbouring locations in Sri Lanka or in southern India. Thus, the war was a period of strained ties between migrant and host fishing communities.

In addition, the tsunami of 2004 negatively affected migrant and host community relations, as a large number of fishers along the northeast and eastern coastal belt lost their lives, homes, and fishing gear, which caused further displacement, poverty, and incapacitation. They had to invest in rebuilding their lives and livelihoods, thus increasing their sense of ownership of the land and its natural resources. The tsunami also resulted in the influx of arbitrarily distributed livelihood support from local and international aid organizations in the form of fishing boats and fishing gear, which encouraged people from non-fishing backgrounds to espouse fishing as a livelihood and which led to increased numbers of fishing households in Trincomalee District.

Thus, in recent years, since the end of the war, when migrants from the west coast have started returning or going for the first time to the northeast and east coasts in large numbers, conflicts between the migrant and host groups have been observed (Lokuge, 2017). Host communities have a stronger sense of regionalism in relation to their natural resources, whereas migrant communities assert their rights to those resources, on the basis of the customary migration patterns of their ancestors and/or the rights of all Sri Lankans to live and work wherever they choose. The conflict has been exacerbated by migrant fishers' use of advanced fishing technology (long lines and larger numbers of gill nets) in contrast to the host communities' use of less-sophisticated fishing methods (smaller numbers of gill nets and purse seines), leading to resource competition and both groups' perceptions that the technology used by the 'other' group puts the sustainability of the resource base at risk.

Furthermore, although the war caused multiple vulnerabilities for fishing families on the east coast, other postwar vulnerabilities have emerged due to resource depletion, climate change, regulatory regimes, and uncertain political conditions, which have contributed to conflicts among fishers. The situation is exacerbated by tourism and industrial development, which has led to overcrowding at beaches and conflict over land between developers and fishers, as also evident in Batticaloa District in Sri Lanka (Chapter 5, this volume) and in Cambodia (Chapter 3, this volume).

Contestation between host and migrant fishers takes the form of sporadic protests and demonstrations, which have taken place in Trincomalee District against migrant fishers, occasionally leading to violent conflict, including the burning of fishing camps, boats, and fishing gear, as indicated in the qualitative interviews. The district fisheries department in Trincomalee has issued decrees to restrict the number of fishing vessels originating from elsewhere, and in neighbouring Mullaitivu District, migrants must work either on the boats of local owners or for beach seine licence holders. Fishers in Mullaitivu have also engaged in disputes with Muslim fishers from Selippur village in Trincomalee District over the use of motorized boats and illegal nets in Kokkilai lagoon, which is accessed by fishing communities from both districts.

Thus, seasonal migration can contribute to new conflicts between migrants and host communities but can also exacerbate ongoing conflicts among local communities, already expressed in ethnic terms. Even conflicts currently perceived as relating to overcrowding and/or resources can easily be ethnicized for political purposes, as communities have been polarized in terms of ethnicity, religion, caste, and politics in Sri Lanka since the country gained independence in 1948. However, as Siriwardena-de Zoysa (2018) points out, in everyday life, conflict exists alongside cooperation for fishing groups, and conflicts within and between local villages are far more common than conflicts between migrant and host communities. Local fishers and migrants continue to fish alongside one another and help one another in times of crisis at sea. Traders from host communities count on migrants to supply adequate fish during the high season, and poor and/or marginalized members of households in host communities rely on migrants to receive fish in exchange for assistance in pulling boats out of the water.

Seasonal migration from the west coast to the east coast and host–migrant relations are matters of substantial debate and frustration for fishing communities in both Puttalam and Trincomalee. While some study participants in the two districts articulated the current conflict in ethnic terms, others expressed it in regional terms of 'north' and 'south', although the communities concerned are located geographically in the northeast and east and in the west respectively. The fact that migrant fishers from some communities are Tamil Hindus of the same caste group (Karaiyar) and others are bilingual Sinhala-Tamil Catholics of the same caste group (Karaiyar in Tamil, Karava in Sinhalese) does not appear to confer any advantages on them in relation to Tamil and Muslim host communities, which regard all migrants as 'southern fishers' and outsiders. Currently, fishers' representatives perceive that they are unable to solve issues relating to migration through discussion and expect the government of Sri Lanka to intervene in the matter. There are several discourses on the issue of migrant fishers in the study sites, and each community considers itself an aggrieved party in the conflict. The debate is framed mainly around the perceived 'right to a tradition of migration' in the study villages in Puttalam and 'the right to one's own resources' in the study villages in Trincomalee.

## Conflicting perceptions: the right to migrate versus the right to local resources

The qualitative data revealed that a combination of material, relational, and subjective well-being factors, such as sustainable income, support from social networks, and a tradition of mobility, were perceived as underlying motivations for internal seasonal migration for both men and women on the west coast. By contrast, host communities' resistance to migrants was perceived in material and relational well-being terms, such as the need for a sustainable income for local people, overcrowding at beaches, and rights to the local fish resources. Thus, the conflict between migrant and host fishing communities seems to stem from a contestation between two overarching discourses on access to fish resources: the migrant fishers' claim to the right to continue the traditional practice of migration, and the host fishers' claim to the right to their local resources. Notions of generational migration patterns; ancestral graves in migratory locations; superior fishing skills; local knowledge of seas, waves, winds, and fish movements; and the socialization of children in migratory locations are part of the discourse of 'the right to a tradition of migration':

> Fishing there [on the northeast and east coast] is a traditional right for us, from generation to generation. The government accepts that. We go there with the support of some people in those villages. There are others who are not entirely opposed to us but they have to show unity by joining with their village and the people who protest that we catch their fish, even though they personally have good relations with us.
>
> (Participant in men's FGD, Kadalpalli, Puttalam)[8]

> My grandfather's grave is in Nayaru. We have had close relationships with Nayaru for generations. There are people in our village who also own land titles [deeds] in Nayaru. There are also villagers who were born in Nayaru, while there are also those born in Selippur as well.
>
> (Wife of fisher, Kadalpalli, Puttalam)

The discourse of 'the right to one's local resources' pivots around deprivation and fishing bans during the war, recovery from displacement, loss of lives, homes and assets, boats overcrowding in areas, and a lack of alternative livelihoods:

> There aren't any big rich people here. There are only people who work for their daily expenses. All of us were affected. We are all people who

> have struggled. There is not a single person here who hasn't gone to India. 100% of us were affected [by the war].
>
> (Participant in men's FGD, Uppukadal, Trincomalee)

> Fishermen here have problems with the migrants. It is the *mudalali* [traders] who profit from them. There are no controls. Sometimes, they bring 300 boats, so local people have problems. How can we all get a catch? We don't go anywhere else, to Kokkavil [village in Mullaitivu District] or Pudavaikattu [village in Trincomalee District], for fishing. This sea has only space for 300 boats. If there are 1000 boats, how can we fish?
>
> (Participant in men's FGD, Selippur, Trincomalee)

Central to the conflict is the shared perception of resource depletion held by both migrant and host communities, entangled in competing claims over the same resource pool. The most recent assessment of fish resources in the coastal waters of the island, marine fish catch statistics at the macro-level, and numbers of fishers in the study villages indicated resource depletion as a critical concern (see Endnote 1). In both the west coast villages and Uppukadal on the east coast, almost everyone indicated that there had been fish resource depletion during their lifetime and within the five years preceding 2017, when the interviews were held. In Kadalpalli, fishers estimated the size of the catch had fallen by as much as 90% and in Uppukadal as much as 75%:

> We think that the fish catch has reduced by 90%. I can remember our father said that it is more than enough to use only six pieces of nets [six whole nets] for a good fish harvest, but today it is not enough even with 30 pieces of nets.
>
> (Participant in men's FGD, Kadalpalli, Puttalam)

> There is a clear decrease in the fish resources. I can say 75% of fish resources have been depleted.
>
> (Fisher, Uppukadal, Trincomalee)

As a result, fishers have to travel farther out to sea to catch fish and use a larger number of nets than they had to previously. Women in all three communities (i.e. the study villages Kadalpalli, Cattiyur and Uppukadal) similarly referred to the reduction in income due to depletion of fish resources: 'The fish resources have depleted. Those days we had a lot of money. In the last five years our income has reduced' (Participant in women's FGD, Cattiyur, Puttalam).

The perceptions of resource depletion and the increased competition for fisheries resources on the east coast were associated with the use of illegal fishing nets and dynamite, as well as overcrowding due to the increased numbers of fishers, boats, and nets. This has led to accusations by migrant fishers and counter-accusations by host fishers that the 'other group' was responsible for the use of illegal fishing gear and methods:

> We do not have serious conflicts with the host communities there. We are all Tamil-speaking communities and there is no language barrier between us. Some of them engage in illegal fishing methods and they do not want us to go there. This is the main reason for the conflict. . . . The people in Selippur use illegal nets. Surukkuvelai [purse seines] are prohibited. Their attitude is, 'If we cannot catch fish, then you should also not catch fish.' It's a kind of jealousy. They also do illegal fishing in Uppukadal.
>
> (Two participants in men's FGD, Cattiyur, Puttalam)

> Migrants come here from Cattiyur, Negombo, Chilaw [town in Puttalam District], Wennappuwa [town in Puttalam District], Kalpitiya. We call them all Cattiyur people. We don't have good relations with them. Business people bring them here. They [migrants] use illegal fishing methods such as *disco valai* (trammel nets). We use only 2.5-inch *thundel* (hook and line).
>
> (Participant in men's FGD, Uppukadal, Trincomalee)

Some fishers attributed resource depletion to poaching by Indian trawlers inside Sri Lankan waters. Moreover, uncertainty relating to fish catches was attributed to climate change in all four study villages. Thus, under the conditions of decreasing and/or uncertain incomes from fishing, fishers in host communities increasingly claim fish resources as their own. Their claim is further strengthened by the threat posed by the migrant fishers' use of more sophisticated fishing methods (long lines and larger numbers of gill nets), whose catch volumes are higher than those of the host villagers, who use other methods (i.e. smaller numbers of gill nets and purse seines). Additionally, the use of the different methods by migrant fishers and host fishers creates the impression that the fishing methods result in the depletion of fish resources, when in practice, migrant and host fishers may be catching different species of fish.

In addition to a shared perception of resource depletion, there is competition for space on the beaches, due to the overcrowding of landing sites and beaches as a consequence of the large influx of migrants who claim a particular location on the beach as a traditional right. Similarly, the overcrowding of boats was evident in Batticaloa District in Sri Lanka (see Chapter 5, this volume). Conflict over landing sites for boats and for constructing *wadis* (camps) sometimes has violent repercussions. When mobilized through fisheries societies, host fishers have organized at district level and actively prevented migrants from establishing camps in their village, or they have negotiated mechanisms (e.g. for imposing boat quotas) with local authorities and politicians. In response, migrants organize to arrive at the host locations on the same day, thereby showing strength in numbers. Fisheries societies in Kadalpalli and Cattiyur have united across ethnic lines to demonstrate against the transport and sale of fish from Trincomalee at the Colombo market if host community fishers prevent them from migrating to the east coast. The hostilities have led to violence, such as the burning of boats and *wadis*, which not only incapacitates migrant fishers, who have to reinvest in their livelihood, but also adds tension and fear to their experience of migration. However, such violent incidents are sporadic and interspersed with longer periods of coexistence and/or cooperation, as argued by Siriwardena-de Zoysa (2018). Moreover, east coast fishers are fully cognizant of how the conflicts are politically constructed:

> We do not have personal conflict with migrants within the *wadi*. Conflicts usually erupt at sea. For example, if a place can accommodate 20 boats, and there are more than that, one person tries to push the other. If there is a problem, we use regionalism. How dare you come to our place?
>
> (Participant in men's FGD, Selippur, Trincomalee)

While disputes and conflicts appear to be manifested over the material dimension of resource scarcity, they are driven by the relational dimension of contestation over access to the fish resources, embedded in the discourses of tradition, suffering, loss, and fairness between the two groups – migrants and hosts. Counter-accusations of the illegal use of fishing gear and constructions of own community as the aggrieved party are articulated in these discourses. Thus, local notions of inequality and perceived marginality (Fabinyi et al., 2015) among west coast and east coast communities and among migrant fishers and local fishers are critical for unravelling these contestations, which underlie conflict.

## The regulatory environment and conflict

Fisheries conflicts are often shaped by the presence and/or intervention of state authorities, which monitor and regulate resource use (Adhuri, 2009). This has consequences for the material and relational well-being of migrant and host communities. The state institutions with regulatory functions over livelihoods in fishing communities are the Department of Fisheries and Aquatic Resources (in the Ministry of Fisheries and Aquatic Resources Development), Provincial Council, District Secretariat, Coast Conservation Department, Department of Agriculture, Department of Irrigation, the police, the coast guard, and the navy. Of these, study participants in the fishing communities were most familiar with the Department of Fisheries and Aquatic Resources. Most participants indicated that officers communicated on fisheries regulations, such as bans on small-size nets, purse seines, dynamite, and lights and the prohibited capture of endangered species, through their community representatives in fisheries societies, among both migrant and host communities. However, individual owners are responsible for obtaining licences annually for their fishing vessels and beach seines. Men in the west coast villages said their relations with officers from the Department of Fisheries and Aquatic Resources were cordial. However, they indicated that the department often did not have the capacity to enforce their regulations, because of pressure from politicians: 'If there's an illegal method being used, there are always some ministers behind that. The fisheries officers also know that but can't do much about it' (Male fish trader, Kadalpalli, Puttalam).

West coast fishers considered that the Department of Fisheries and Aquatic Resources was unable to prevent violations of regulations by east coast fishers, while many fishers in Selippur and Uppukadal perceived that fisheries officers were partial and enforced their bans only on local fishers on the east coast. Muslim fishers indicated that both Sinhalese fishers from the west coast and Tamil fishers were shown partiality by the state authorities: 'Government officers are biased. . . . They catch only boats [using banned nets] of Muslims but not boats of the Tamils and Sinhalese' (Participants in men's FGD, Selippur, Trincomalee). Tamil fishers in Uppukadal claimed that when migrants faced conflicts, such as over securing spaces for their boats, they received the support of fisheries authorities with regard to using banned nets and the support of the police:

> The only problem is the migrant fishers. If you use a three-quarter inch to one-and-a-half inch eye net [1.9–3.87 cm mesh net] and cast it on the rocks, the next day there will be no fish there. If you cast nets on rocks, the fish roe will also get trapped, so the fish resources decrease. Sea plants also get destroyed, so fish don't have a place to lay eggs. Therefore, there are no fish. The Fisheries Minister will be from their area. The second problem is that they [fisheries authorities] view Tamils as second-class citizens.
>
> (Participant in men's FGD, Uppukadal, Trincomalee)

> There has been so much violence both on the shore and the sea. We don't talk to them [migrant fishers]. If we have disputes and go to the police, they just say 'Go, go!' Whether the police officers are Tamil, Muslim or Sinhalese, the migrants have political support.
>
> (Participant in men's FGD, Uppukadal, Trincomalee)

In response to local protests and political intervention, the Department of Fisheries and Aquatic Resources in Sri Lanka has imposed quotas on the number of migrants who can bring their own boats to fish from landing sites on the northeast and east coasts. The entire Trincomalee District has a quota of 300 migrant boats per fishing season (fisheries officer, Trincomalee

District); the quotas for Selippur are 117 migratory boats and those for Uppukadal are 49 migratory boats, per fishing season, respectively. According to a representative of the fisheries authorities, the quotas had been observed in 2016 and 2017 (fisheries officer, Trincomalee District). The authorities issue permits for nets and check on the licenses for the boats, communicating with fishers through their societies. Nevertheless, both migrant and local fishers claimed that the numbers of migrant boats, despite having reduced in number in the two years before the interviews in 2017, exceeded the quotas set by the Department of Fisheries and Aquatic Resources.

It appears that rent-seeking behaviour framed by institutionalized ethnocentrism leads to different levels of access to migratory sites for different stakeholders. The strong social networks of migrant fishers enable some to circumvent state restrictions due to connections with politicians, fisheries officials, and church authorities within the ethno-political structure of the country. Additionally, host fishers alleged that migrant fishers were shown partiality by the police and navy due to their bilingualism and assertion of their majority ethnic identity.

An important yet indirect agent in gaining access to migratory sites are local *mudalali* (beach seine and boat owners and fish traders) from the east coast, who recruit workers from the west coast to fish with their beach seines or on their boats and/or guarantee the purchase of fish during the fishing season. These agents often have great influence in the host village and/or district and are able to negotiate with authorities and regulatory bodies on behalf of the migrant fishers and workers:

> Sinhalese and Muslim *mudalali* (traders) bring migrants here, not our [Tamil] *mudalali*, who are concerned about local fishermen. Business people think that only they should live. We think all of us should live. If our society decides to sell a kilo of fish for LKR100 [USD0.57], the migrants sell for LKR60 [USD0.34]. Migrants have earned lakhs of rupees [hundreds of thousands of rupees] from the *mudalali* who have brought them here.
>
> (Participant in men's FGD, Uppukadal, Trincomalee)

Thus, from the east coast fishers' perspective, the material well-being of local *mudalali* is achieved at the cost of the material well-being of their own community. While *mudalali* are perceived to enhance the material well-being of migrants, it is at the cost of relational well-being of host and migrant fishing communities

Fishers in Selippur indicated that they often faced the confiscation of their nets (small mesh-size gill nets and purse seines) and boats and faced arrests, and they had to spend large amounts of money to have their fishing vessels, fishing gear, and themselves released. Moreover, whereas both villages on the west coast and Uppukadal on the east coast expressed a significant depletion in fish resources compared to the past, the Muslim fishers of Selippur

35 km north of Uppukadal expressed a contradictory opinion, namely that fish resources in the sea and Kokkilai lagoon were abundant and sustainable. They explained this as being due to the will of God, who provided as the human population increased. Behind their contradictory perspective lay not only the religious belief that God would provide but also a challenge to fisheries authorities' regulations and restrictions on mesh sizes and fishing methods (such as purse seines, lights, and dynamite), which affect the practices of fishers, who have customarily caught small fish in Selippur by using these currently banned methods. The claim by fishers from Selippur that fish resources were not depleted therefore appears to have called into question the imposed restrictions. Most fishers from Selippur indicated that they responded to bans by continuing to use their fishing methods while escaping the scrutiny of the authorities and paying to recover their boats or fishing gear whenever they were confiscated. Some said they planned to start using alternative, permitted fishing gear.

The enforcement of bans on specific fishing gear has led to conflicts not only between government authorities and fishers from Selippur but also between the two fishing communities on the east coast. The conflict between fishers and government authorities is linked to a lack of transparency regarding the imposition and rationale behind fisheries bans and regulations, the failure of fisheries authorities to provide information and resources on alternate methods of fishing, delays in issuing licences for the seasonal use of purse seines, and the general climate of suspicion of the state's purported ethnocentric agenda in the east. By contrast, the tensions between fishers from Selippur and Uppukadal stem from accusations by Uppukadal fishers that those from Selippur continue to use illegal fishing methods. This also takes an ethnic turn, where Muslims are accused of using illegal fishing methods, countered by fishers in Selippur who claim that both Sinhalese and Tamils receive the favour of the fisheries authorities. However, the conflict between state authorities (fisheries officers and navy officers) and fishers on the east coast over fishing gear appears to be more pronounced than disputes among fishing communities on the east coast. Perceptions of rights to migration and local resources, which emerged in the discourses of migrants and host groups respectively, are enmeshed with perceptions of inequality and fairness on the part of regulatory authorities in their relationships with the two groups. Horizontal conflicts among fishing communities that are based on local perceptions of inequality (Fabinyi et al., 2015) can be aggravated by vertical conflicts between fishing communities and regulatory authorities over access to the fish resources and control of fishing gear, thereby increasing the tendency for the conflicts to be ethnicized or regionalized. Thus, state regulatory institutions contribute to exacerbating conflicts over the fisheries resources on the east coast, both reinforcing the north–south binary, which emerged in the postwar period between migrant and host fishers, and increasing the potential for ethnicizing this conflict.

## Conclusions

In this chapter, we have aimed to understand the underlying sociopolitical and environmental causes of contestation, disputes, and conflict between migrant fishing communities on the west coast and host fishing communities on the east coast of Sri Lanka. Seen through a social well-being lens, the motivations underlying the conflict between migrant and host groups is revealed in two competing discourses on access to resources in the pursuit of well-being: the right to continue a tradition of migration and the right to one's local resources. Although disputes and conflict are manifested over the material dimension of fisheries resources, they are driven mainly by the relational dimension of contestation, encompassed in the discourses of the rights to access those resources, in terms of local inequalities and perceived marginality, as argued by Fabinyi et al. (2015).

Both migrants and host groups seek similar forms of material well-being, such as increased and sustainable incomes through the fisheries resources. However, the material well-being outcomes for migrants are higher in terms of their returns from fishing, because their assets and skills levels are higher. By contrast, host fishers, impoverished by war, possess less-sophisticated fishing vessels and fishing gear, and they have fewer skills, as many are newcomers to fishing. Both groups are affected by depleting fish resources. Stronger social networks of migrants provide higher relational well-being outcomes in terms of institutional support to access fisheries resources, although this is at the cost of amicable social relations between migrant and host groups. The subjective well-being of migrants is linked to the value attached to mobility and to a sense of emotional attachment to the migratory location in a culture of migration. With increasing frustration and anxiety over potential hostilities between migrant and host communities, the subjective well-being of both groups appears to have decreased.

From a political ecology perspective, state institutions both contribute to and exacerbate resource conflict between the migrant and host groups by imposing non-transparent regulations, which show little understanding of the socioeconomic impacts on specific communities and are therefore perceived as favouring migrant fishers over host fishers. This strengthens the north–south divide that has emerged among these groups after the civil war in Sri Lanka.

## Notes

1 The figures for 1983 and 2005 are from an unpublished Ministry of Fisheries and Aquatic Resources Development annual report on fisheries statistics dated 2015.
2 In Sri Lankan fisheries reports and statistics, the abbreviation OFRP (outboard-engine fibreglass reinforced plastic) boat is commonly used to refer to small motorized fibre-reinforced plastic boats.
3 In Sri Lankan fisheries reports and statistics, the abbreviation NTRB (non-motorized traditional boat) is used to refer to traditional canoes and rafts.

4 Personal communication in 2019 from the National Aquatic Resources Research and Development Agency (NARA) on the preliminary findings of the Survey of Regional Resources and Ecosystem off Bay of Bengal: Sri Lanka)
5 Data were derived from a household survey, focus group discussions, preliminary livelihood profiling, community histories and migration histories, in-depth open-ended household interviews, key informant interviews, and field observations.
6 The village names are pseudonyms, in accordance with ethnographic convention.
7 R.L. Stirrat, 1972, notes on Ceylon Administration Reports, 1867–1887, held in personal archive material
8 The quotes from study participants in Kadalpalli village were translated from Sinhala into English, and all other quotes from study participants were translated from Tamil into English. The translations were done by the researchers who conducted the interviews, among whom were the authors of this chapter.

## References

Abobi, S.M., and Alhassan, E.H. (2015). A review of fisheries-related human migration in the Gulf of Guinea. *Journal of Coastal Zone Management*, *18*, 395. doi:10.4172/2473-3350.1000395

Adhuri, D.S. (2009). Social identity and access to natural resources: Ethnicity and regionalism from a maritime perspective. In M. Sakai, G. Banks, and J.H. Walker (Eds.), *The politics of the periphery in Indonesia: Social and geographical perspectives* (pp. 134–152). Singapore: NUS Press.

Amarasinghe, O. (1989). Technological change, transformation of risks and patronage relations in a fishing community of South Sri Lanka. *Development and Change*, *20*(4), 684–734.

Bavinck, M., Pelligrini, L., and Mostert, E. (2014). Introduction. In M. Bavinck, L. Pelligrini, and E. Mostert (Eds.), *Conflicts over natural resources in the Global South: Conceptual approaches* (pp. 1–11). Boca Raton, FL: CRC.

Bennett, N.J. (2019). In political seas: Engaging with political ecology in the ocean and coastal environment. *Coastal Management*, *47*(1), 67–87. Retrieved from https://doi.org/10.1080/08920753.2019.1540905 accessed 19/12/2019.

Fabinyi, M., Foale, S., and MacIntyre, M. (2015). Managing inequality or managing stocks? An ethnographic perspective on the governance of small-scale fisheries. *Fish and Fisheries*, *16*, 471–485.

Frerks, G., Dietz, T., and van der Zaag, P. (2014). Conflict and cooperation in natural resources: Justifying the CoCooN Programme. In M. Bavinck, L. Pelligrini, and E. Mostert (Eds.), *Conflicts over natural resources in the Global South: Conceptual approaches* (pp. 13–34). Boca Raton, FL: CRC.

Jul-Larsen, E. (1994). *Migrant fishermen in Congo: Tradition and modernity*. CMI Report No.6. Bergen: Chr. Michelsen Institute.

Lokuge, G. (2017). *Even fish have an ethnicity: Livelihoods and identities of men and in war-affected coastal Trincomalee, Sri Lanka*. Doctoral thesis, Wageningen University, Wageningen.

McGregor, A. (2008). *Well-being, poverty and conflict*. ESRC Research Group on Well-being in Developing Countries Briefing Paper 1/08. Bath: University of Bath.

Ministry of Fisheries and Aquatic Resources Development. (n.d.). *Fisheries statistics 2017 [2016]*. Retrieved from www.fisheries.gov.lk/web/images/downloads/pdfs/fisheries_statistics_2017.pdf accessed 19/12/2019.

National Aquatic Resources Research and Development Agency [NARA]. (n.d.). *Sri Lanka fisheries yearbook 2003–2005*. Retrieved from www.erepository.nara.ac.lk/bitstream/handle/1/1188/NR-SOCIO-19-FULL-TEXT.pdf?sequence=2&isAllowed=y accessed 19/12/2019.

Roberts, M. (1982). *Caste conflict and elite formation: The rise of a Karava elite in Sri Lanka, 1500–1931*. Cambridge: Cambridge University Press.

Salayo, N., Ahmed, M., Garces, L., and Viswanathan, K.K. (2006). An overview of fisheries conflicts in South and Southeast Asia: Recommendations, challenges and directions. *NAGA WorldFish Center Quarterly*, *29*(1–2), 11–20.

Siriwardena-de Zoysa, R. (2018). *Fishing, mobility and settlerhood: Coastal socialities in postwar Sri Lanka*. Cham: Springer International.

# 5 Female headship and exclusion from small-scale fishing in Eastern Province, Sri Lanka

*Ragnhild Lund and Fazeeha Azmi*

## Introduction

This chapter investigates social exclusion in small-scale fishing communities in Ampara District and Batticaloa District (Eastern Province, Sri Lanka) with respect to vulnerable female household heads (FHHs). Rather than compare male household heads and female household heads, our intention is to explain how, in situations of displacement due to previous war and tsunami, FHHs are particularly vulnerable and excluded. We explore the specific attributes of the head that make some female-headed households more or less vulnerable (Beegle and van de Walle, 2019).

In the following, we attempt to understand the situation of FHHs through the lens of social exclusion and ask how mobility and migration are both causes and effects of exclusionary processes among individual fishers and their communities. This topic derives from our previous work on the social well-being of FHHs in southern India (Azmi et al., 2020) and in eastern Sri Lanka[1] for the research project Migration and Collectives/Networks as Pathways out of Poverty: Gendered Vulnerabilities and Capabilities amongst Poor Fishing Communities in Asia. We found that vulnerability and well-being was intertwined with exclusion and precarity. In the study conducted in India, many FHHs faced a precarious life situation because they were prevented from fishing because of mechanization and deep-sea fishing (Chapter 2, this volume). We found that their subjective well-being perspectives overlapped their material and relational well-being, because the women emphasized issues related to their employment and related to their family members' employment, their children's education, and pensions. When FHHs were the sole income earners in the family, they did not have any money left to invest in housing repairs, clean water, or savings, which we identified as important dimensions of well-being. In the Sri Lankan situation, the implications of the civil war and the 2004 tsunami have continuing effects on fishers' livelihoods, and the well-being of female fishers has become more vulnerable despite their various strategies to achieve their well-being targets.

There are many reasons and routes for women to assume the headship of a household, which thus results in inconsistencies and complexities in identifying who female heads of the households are (Buvinić and Gupta, 1997; Chant, 2004; Ruwanpura and Humphries, 2004). Ruwanpura and Humphries (2004) explore the experience of FHHs in Eastern Province in Sri Lanka and point out how their differential routes to becoming FHHs result in various survival strategies. The authors document how thousands of women were both forced to become FHHs after the civil war and the 2004 tsunami and forced to abandon their traditional women' role in order to engage in income-earning activities. The women had to negotiate their new roles in the highly patriarchal context underpinned by strong sociocultural practices that influenced their mobility, livelihood choices, access to resources, and access to markets (Ruwanpura and Humphries, 2004). In these spaces, FHHs may face exclusion. With one exception, all FHHs in our Sri Lankan study were either the sole contributor or the main economic contributor in their household, depending on whether they were widowed, divorced, separated, abandoned, or unmarried.

People living on the east coast of Sri Lanka suffered from the effects of the 2004 tsunami, when they had to flee from the villages that were closest to the sea. After the tsunami, a buffer zone was created on the affected coastline, and people were displaced to the interior of the country and often far away from their traditional villages. Currently, some of the displaced people are still staying in the new settlements. Many of them do not have regular income. Others have migrated to other countries, particularly to the Middle East, because they found it hard to survive in the new settlements and to return to fishing for a livelihood. To the north of Batticaloa town, people were active in fighting the war, which in Eastern Province ended in 2007, and today people generally find it hard to find new livelihoods or return to fishing. In their case too, overseas migration has been sought as a way to overcome their vulnerability.

Currently, Eastern Province is ranked the second poorest province in Sri Lanka (Department of Census and Statistics, 2017). Whereas Batticaloa District ranks first in terms of the poverty head count index (Department of Census and Statistics, 2017), Ampara District ranks fourth (Department of Census and Statistics, 2017). It is pertinent to point out that Batticaloa District also ranks first with respect to the percentage of FHHs, which is 32.3%, whereas the percentage of FHHs in Ampara District is 24% (Department of Census and Statistics, 2018). Thus, many of the female-headed households in the two districts are vulnerable and, by implication, socially excluded. Multidimensional disadvantages prevail in small-scale fishing communities, such as a lack of steady income from fishing, lack of schooling, lack of health facilities, and lack of proper housing.

In the fisheries sector, vulnerability is identified as an important factor in fisher's poverty (Béné and Friend, 2009). The vulnerability of fishers can be attributed to physical, socioeconomic, and political factors that influence an individual's or a group's ability to withstand unfavourable conditions (Béné and Friend, 2009; Coulthard, 2008). Recent research on vulnerability and precarity in the Global South has distinguished between inherited vulnerability (low class, caste, and identity) and produced precarity. In this chapter, we regard *vulnerability* as associated with factors such as displacement and disrupted families, whereas *precarity* refers to people whose livelihoods are rendered unstable and insecure due to recent modernization programmes, land grabbing, and a lack of savings from migration overseas in the postwar and post-tsunami period.

To unravel vulnerability and precarity in this chapter, we first discuss the term *social exclusion*. Rather than counter-positioning *exclusion* and *inclusion*, our intention is to show how and why FHHs are excluded and with what consequences. Thereafter, we describe the role of small-scale fishing in Eastern Province, followed by a brief presentation of our methods. Drawing on participants' stories collected at three locations in Eastern Province, we seek to address the questions of the role of fishing in vulnerable situations for FHHs and whether migration serves as an alternative solution to their problems. In contrast to Chapters 4 and 6, which deal with west coast to east coast migration in Sri Lanka, this chapter deals with displacement and overseas migration. We end the chapter with a concluding discussion.

## Social exclusion

Social exclusion is the process whereby individuals or groups are wholly or partially excluded from full participation in the society in which they live (de Haan, 1998; Francis, 2002; Rawal, 2008). An example of individual exclusion is the exclusion of single women from meaningful work; other forms of social marginalization may occur due to ethnicity, marital status, gender, or class. Furthermore, entire communities may experience social exclusion, often as a result of displacement or minority status.

Many scholars consider social exclusion multidimensional, on the basis that it captures the multifaceted nature of social deprivation, especially in its institutional and cultural aspects (Aasland and Fløtten, 2001; Benington, 2001; Francis, 2002; O'Reilly, 2005; Silver, 1994, 2007). Such scholars broaden the notion of material poverty and identify social problems, labelling them aspects of social exclusion. This view of exclusion is in accordance with that of Sen (2004), who argues that social exclusion must be examined in relation to its utility in providing new insights into the nature and causes of poverty and furthering our thinking about policies and social action to alleviate poverty. Hence, social exclusion extends beyond material poverty, as it incorporates social disadvantages, such as access to education, healthcare, social care, and housing. In such a view, social exclusion is not only the outcome of a process but also a process in itself. Hence, it is dynamic rather than static, and therefore, people may experience different stages of social exclusion at different points in their lives, which in turn may affect their well-being.

Furthermore, social exclusion is context-specific (INCLUSO, n.d.). Hall et al. (2011) have addressed how any productive use of land requires the exclusion of some potential users and found that most projects involving transformed land relations are accompanied by painful dilemmas with respect to identity and belonging. They argue that rather than counter-positioning *exclusion* and *inclusion*, attention must be paid to who is excluded, how, why, and with what consequences. Moreover, social exclusion can be understood from an individual or a collective perspective (INCLUSO, n.d.). From the individual perspective, social exclusion entails a lack of access to resources and opportunities. From a collective perspective, it refers to social problems that divide the majority from the minority of marginalized, unemployed, or 'aggrieved human beings' (Sen, 2000).

Lastly, social exclusion is relational in that it involves both excluders and those who are excluded (Cameron, 2006; INCLUSO, n.d.). This calls attention to how inclusion can be obtained through better policies and practices. In such a perspective, exclusion and inclusion are two sides of the same coin. Only by addressing the question of what constitutes exclusion can the question of what constitutes inclusion be addressed; each question is mutually dependent on the other (O'Reilly, 2005, p. 84).

## Small-scale fishing communities in Eastern Province

Eastern Province in Sri Lanka has a long history of fishing that has supported the livelihoods of many coastal communities and people living in Batticaloa town, which is known as 'the land of singing fish' due to the abundance of fish. In these places, fishing has traditionally occurred in rivers, lagoons, the near shore, and the deep sea, from April to October, with an off-season during the northeast monsoon in Asia (between November and January), which brings strong winds and rough seas.

The armed conflict between the government of Sri Lanka and the Liberation Tigers of Tamil Eelam (LTTE) (1983–2007 in Eastern Province) seriously disturbed and disrupted everyday life in Eastern Province and access to the sea and coastal areas, which significantly affected the economic activities of coastal communities (Lehman, 2013). For reasons of security, fishing activities in the province were disturbed, with differing temporal and spatial intensities between the years 1983 and 2004. In particular, the situation affected a large number of small-scale fishers who depended on the local fish, sea, and coast for their livelihoods (De Silva and Masahiro Yamao, 2007). Today, fishing continues despite the former difficulties. However, in certain locations, it is quickly disappearing as a main way of life.

With the end of the war in Eastern Province in 2007, coastal communities that had been displaced due to the war (and, before that, the 2004 tsunami) began to resettle in their former villages or in new fishing settlements. However, new challenges were created by the postwar and post-tsunami situations. In particular, access to the sea and coast was restricted due to unclear regulations regarding arbitrary buffer zones, environmental changes, and postwar development activities. According to the latest available statistics from the Ministry of Fisheries and Aquatic Resources Development (Ministry of Fisheries and Aquatic Resources Development and Rural Economy, 2018), fish catches in the study districts (Ampara and Batticaloa) have remained low, while the number of households engaged in fishing has increased. In the absence of other livelihood options, many people are turning to fishing, and some of these new fishers are using destructive fishing methods that are resulting in declining fisheries resources. Apart from this situation, the new fisheries policy of Sri Lanka – which, for economic reasons, encourages mechanized deep-sea fishing practices – is likely to have a negative impact on the lives and livelihoods of many small-scale fishers who currently occupy coastal settlements. An interview with a fisheries officer in one of the study districts revealed that near-shore and lagoon fisheries will soon be restricted due to ecological and economic considerations. It follows that such actions will affect the fisheries-based livelihoods of many small-scale fishers and particularly FHHs who engage in difficult and risky fisheries-related activities. This scenario applies also to the study locations.

In eastern Sri Lanka, it is common for women from poor families and FHHs to be involved in fishing, particularly in midlife or old age. The women we studied in Batticaloa District were directly involved in fish processing and selling (the poorest) or indirectly involved in sorting fish, mending nets, and drying and smoking fish (the more prosperous). The women who sold fish at the market were in a relatively better position than the women who sold mussels at the market, at home, and on the sides of the road (the very poor). In Ampara District, the women engaged in different income-earning activities, but they were originally from fishing communities. Many of the women from both districts were previously migrant workers in the Middle East. When they were abroad, they were unable to save for their return, because the costs of living in the host country were high and because women generally had menial jobs. Hence, when they returned, they were particularly vulnerable and faced a precarious situation. In addition, several of the older fish vendors faced the burden of caring for grandchildren, whose parents were abroad.

## Methods

Our fieldwork took place in Ampara District and Batticaloa District in Eastern Province. Having done fieldwork earlier in these districts (2012–2017), we were familiar with the socioeconomic situation (Attanapola et al., 2013; Azmi et al., 2013; Blaikie and Lund, 2010) and randomly chose the three study areas for our revisit, one urban (in Batticaloa District) and two rural (in Batticaloa and Ampara Districts). We chose to study FHHs who worked in and around Batticaloa town and FHHs who lived in two rural-based

*Table 5.1* Details of the study participants in Batticaloa District and Ampara District, Sri Lanka, February 2018

| *Name (pseudonym)* | *Age (years)* | *Ethnicity* | *FHH* status* | *Dependents in household* | *Work* | *Mobility and migration* |
|---|---|---|---|---|---|---|
| Sarojini | 38 | Tamil | Joint provider (main provider in Batticaloa) | 2 daughters | Fish vendor | Travelled daily to fish market; husband had migrated to Middle East |
| Kamala | 52 | Tamil | Sole provider (husband unemployed) | husband, 1 daughter, 1 granddaughter | Gleaner and vendor | Sold mussels from home |
| Fathima | 52 | Muslim | Sole provider (abandoned by husband) | 3 sons (one married, one given away) | Mat weaver (occasional) | Migrated to the Middle East; son migrated to the Middle East |
| Sakeena | 39 | Muslim | Sole provider (widow) | 2 daughters, 3 sons | Made food to sell | Displaced; sold food from home |
| Ayesha | 57 | Muslim | Sole provider (widow) | 3 sons (one married, two in the Middle East) | Unemployed | Displaced; previous migrant to the Middle East; sons migrated to the Middle East |
| Ponnamma | 60 | Tamil | Sole provider (husband unemployed) | 1 son, 2 daughters (all deceased); looked after 3 grandchildren (two boys, one girl) | Fisher | Postwar returnee |
| Gowry | 60 | Tamil | Main provider (widow) | 3 daughters | Fisher | Postwar returnee |

Note: *FHH – female head of household
Source: Interview data 2017

areas north and south of the town in order to study spatial and occupational changes in fishers' lives as perceived by the FHHs.

We conducted the new fieldwork in both of the study districts from 6 to 17 February 2018. During that period, we undertook four group discussions with researchers at Eastern University, Sri Lanka, a local NGO, and the fish market in Batticaloa town. We interviewed a representative of the Ministry of Fisheries; fishers and fish vendors at Batticaloa fish market and in the surrounding villages (12); and fishers directly affected by tsunami-induced and development-induced displacement farther south in Ampara District (3). FHHs were chosen at random. For the interviews, we used an open-ended question guide, which proved a flexible methodological tool. The interviews were translated from the Tamil language into English by one of the coauthors of this chapter and immediately transcribed by her because she is fluent in both languages. In addition, we maintained a separate diary in which we recorded our observations, reflections and photographs, which provided useful information for the subsequent analysis and writing.

In this chapter, we present the stories of seven women (Table 5.1): two who were engaged in fish vending at the market and at home in Batticaloa town, one from a tsunami-affected village, two from a new settlement, and two from a war-affected village. The women were representative of FHHs in the selected study villages and hence did not constitute a representative sample of *all* households. Information gained from their individual stories was supplemented with information gained from other interviews, informal conversations, and observations.

## Fish vending in and around Batticaloa town

Studying the FHHs in and around Batticaloa town informed us about inclusion and exclusion in a hierarchy of fish vendors. The fish vendors we met had earned a decent living and had managed to live well, while others struggled for survival on a day-to-day basis. We learned about the women's success and well-being, as well as their struggles and tribulations.

The fish market in Batticaloa town is in a new, large open-air space. On the first day of the revisit in 2018, we went to the fish market to observe the activities of the fish vendors. Nine female fish vendors were present, five of whom were willing to talk to us. Three of the women had their own stalls. Of those three women, two were in their 40s and married to husbands who worked in Saudi Arabia. The latter two were relatively well off. They owned good houses in Batticaloa town and could afford to buy fish for resale. In the evening, some of the women sold fish at the night market. The third woman was elderly and had sold fish for 30 years after her husband died. She told us that she bought fish for about LKR500 (USD2.3). The other two female vendors took their own fish to the market and sold sprats and small oysters while sitting on the floor. Thus, there was an observable hierarchy of fish vendors, divided between those who were earning a decent living and enjoying an orderly work situation and those who were struggling to survive in a precarious situation.

Sarojini,[2] one of the vendors in the market, lived in a good house, with a garden with fruit and coconut trees, and kept chickens. Inside, her house was spacious and well furnished. Somewhat unconventionally, Sarojini started working at the fish market at the young age of 24. Over the years, her business prospered, and at the time of the interview, she earned between LKR1500 (USD8.48) and LKR2500 (USD14.13) per day. She worked seven days per week from 7:30 a.m. to noon but did not need to sell at the night market. She bought her fish from wholesalers in nearby markets.

Sarojini's husband had various jobs – baking, fishing, painting, and driving – and he lived and worked in the Middle East. He first went to Qatar when Sarojini became pregnant with their second daughter, at which time he worked for a couple of years before returning home. Shortly thereafter, he went to Saudi Arabia, where he remained for almost 16 years (with the exception of some time spent in Kuwait). After years of hard work as a driver, he was able to rent a truck in Saudi Arabia and earn a reasonable income. At the time of our meeting, he was back in Sri Lanka, preparing for the couple's oldest daughter's wedding, but he was returning to Saudi Arabia after the wedding. His planned to build a house for his second eldest daughter as a dowry, and he and Sarojini already had sufficient money to buy a plot of land.

Sarojini and her husband's story informed us of how pooling resources in Sri Lanka and abroad (through migration) enabled them to live fairly well, to ensure the future well-being of their daughters and to allow Sarojini to continue work as a fish vendor. Sarojini was well established in the fish-vending business.

After the war ended in 2008, Kamala, along with her husband and daughter, returned to her home village from a refugee camp. When she was growing up, everyone in her family was a gleaner (for oysters). When Kamala married, she continued in the same line of work, and her husband worked on a big boat. After the war, her husband lost his job, but she continued to fish on a small boat with her brother for some years until he passed away. She said,

> In this village, everybody is involved in fishing. People fish for their food and to make an income by selling fish. During the war, we were stopped by the army and navy, and they threatened us, saying we should not to go to the sea. Some of us did fishing despite those threats because we did not have money to feed our children. . . . During that time, our men were working as coolie workers in paddy lands. I sold my jewellery to buy knives and other implements for agriculture. With the income . . . we were able to feed our families. . . . After that [the war], we came to our villages and found nothing was left at home. . . . So, we had to start our lives from the beginning. After coming to the village, my husband

> started to fish again, but it was difficult to go to sea, as the . . . army was still there. They [people in the village] had to go to the sea secretly.

Kamala walked to the river (approximately a 4 km distance) with three other women, where she gleaned mussels from her brother's boat from 9 a.m. to 3 p.m. daily during the mussel-harvesting season. She did not glean during the rainy season, when the water is cool; instead, she did other types of work in the village, such as cooking school meals. Kamala normally carried the mussels on her head while walking or using her bicycle. The mussels were generally sold for LKR1000–1500 (USD5.65–8.48). She boiled and dried any mussels that were not sold on the same day as they were caught. There was a demand for both fresh and boiled mussels. Kamala did not sell mussels outside her village. Through her work, Kamala was able to feed her family (husband, daughter, and granddaughter), which she experienced as a constant struggle for survival.

## Post-tsunami displacement

In this section, we present the story of Fathima, who lived in an old village in Ampara District, as well as the stories of Sakeena and Ayesha, who lived in a post-tsunami settlement. Their stories informed us of the interlinkages between migration, displacement, and lack of access to fishing grounds.

### *The old village*

The old village is a densely populated Muslim fishing village in Ampara District. Before the 2004 tsunami, the men went fishing and many of the women worked at a local weaving centre. The village was severely damaged by the tsunami: hundreds of people lost their lives, and almost half of the village houses were severely damaged. Moreover, due to the buffer zone that was set up by the government, many villagers were prevented from moving back to their homes. Instead, they were resettled in an area located 50 km inland. With the help of an international NGO, the new settlement provided two-room houses and basic infrastructure, such as a primary school, a road, and a water tank. While some of the local fishers continued to live in their pre-tsunami houses, the majority, whose houses had been washed away or damaged, were relocated to settlements. Fifteen years after the tsunami, many houses remain in a state of disrepair. Furthermore, the construction of a new harbour eroded and subsequently destroyed the beach. Villagers have had to build a sea wall of rubble to prevent the persistently rough sea waves from inundating their homes. In addition, another actor (a government higher education institution) has taken land from the villagers and erected a huge wall to exclude them.

Fathima was born in the old village. She told us that she first went to the Middle East when she was 20 years old and newly married. She moved there 'only to work' and stayed there for almost 10 years. During that time, Fathima's husband was involved in fishing. He left her when the youngest of their three sons was only four months old, and she became responsible for her family. Fathima found her life difficult, to the extent that she was forced to give her youngest son to relatives. At the time of our meeting, she wove mats and sold each of them for LKR1350 (USD7.63).

Marital problems are often caused by migration. Fathima said, 'Women go to the Middle East and send money. When they come back, they find that everything is spent and the husband has not been doing any work. This leads to divorce in many families in our village.' Moving to the Middle East led to many disrupted families in the home village because female migrants stayed away for a long time and could not control the investments at home.

The new developments in the village (the harbour and university) contributed to reducing fishing stocks, thereby leading to further unemployment, particularly among young men. According to the villagers we talked to, fishing was no longer an attractive occupation and did not enable households to have a sustainable lifestyle. It appeared that no one in the village had gained from the harbour construction: the sea levels had risen, and there are no fish close to the beach. Furthermore, fishers could not take their boats out to fish in the deep sea. The problem primarily related to the harbour, followed by insufficient market and storage facilities to keep fish fresh until sold, which rendered fishing no longer profitable. Men take up seasonal work only occasionally, as coolie labourers, but the men we talked with said that they could not save much from such work and that if they fished, their catches were taken by the boat owner. The female fish vendors had lost access to fish, so it had become difficult for them to find work, especially those who were the main providers. Both male and female fishers became redundant: 'nobody is coming back to fishing' (Fathima).

### *The new settlement*

One of the coauthors of this chapter visited the new settlement several times after its construction (2004–2006). At the time of re-visiting the area, her impression was that a large part of the village was empty, with only 52 houses inhabited. The road was almost inaccessible by motorized vehicle, the school looked rundown, the water tank was dry, and there was only one small tea shop in the village. Some men regularly went fishing along the coast next to their old village despite the long journey and difficult access, but more men did seasonal fishing on the coast, during which time they constructed fishing huts along the beach. One house owner in the old village told us that migration, not fishing, sustained the livelihoods of most families in the settlement. For several years, both women and men from the village had migrated to the Middle East for work. However, views on migration varied among the women we interviewed. While some of the women saw overseas migration as a positive option, others saw it as a threat to family unity and local stability.

Sakeena had been a widow for four years. Like Fathima, she was from the old village, where she had lived on a street that ran parallel to the beach, which was the worst damaged street during the tsunami. When she was 13 years old, her parents gave her away to a Muslim family, to work as a maid. The family was kind to Sakeena and treated her as a daughter; at age 15, she converted to Islam and married a relative of the Muslim family. Her husband had worked as a fish labourer for four years before the 2004 tsunami, but after the disaster, they were forced to move to the settlement, where her husband took up work loading sand onto a truck. Sakeena's children, aged between 8 and 20, had been able to attend school due to the financial support of a local philanthropist. Her husband was seriously ill for nine years and could not work, so she started to cook meals for the local schoolchildren. Through this work, Sakeena earned LKR150–200 (USD0.85–1.13) per day. Her life was extremely difficult, but she was able to manage, due to the philanthropist's contribution to her children's schooling (worth LKR10,000 (USD56.55) every three months). Sakeena's main hope for the future was that her children, especially her sons (ages 15, 19, and 20), would fare well. She said,

> I am living for my children. . . . If my sons can find jobs that would be good. . . [but] it is difficult. If they cannot find work here, I will have to send them to the Middle East as a last option. . . . They are boys . . . they can go anywhere to do jobs, not like girls.

From the quote, we understand that Sakeena perceived migration as a last resort and as more acceptable for men than for women.

By contrast, Ayesha saw migration as the only viable option in difficult times. One of her sons worked in Qatar, one was a local labourer, and her youngest son had recently left to work in Qatar. Ayesha's house, which had been given to her by an NGO after the tsunami, was only half built and was in poor condition. The youngest son had left home to help Ayesha build a better house. She had previously worked for 10 years as a housemaid in the Middle East. She had also worked in several countries. Due to the harsh working conditions of her previous employments, she suffered from poor health (asthma):

> My husband was in fishing and he did not know anything but fishing. After we came here, we found it difficult to live and I decided to go. . . . I am sad when I think of our life in our village before the tsunami. We lived close to the beach. Our family was a fisher family, and whatever we earned was enough for food and other needs.

Ayesha emphasized that staying abroad restricted her possibilities to save and plan for the future, because much of the income was required for food and housing overseas. Thus, elderly returnees frequently live in misery:

> Now, although it is nearly 15 years since we came here, we do not have very close friends. . . . Relatives and friends are important in our life. Many of my relatives who came here also went back to the village. . . . They cannot help us.

Ayesha's story told us about a life spent on the move, distanced from traditional fishing culture and social norms, and excluded from friends and relatives. Although she had lived abroad for a long time and worked hard, she had not had a good life. She was starving and suffering from poor health.

## Postwar period

We studied a village located north of Batticaloa town, which had been a strategically important location for both the LTTE and the government during the civil war. The village suffered greatly during the war and 2004 tsunami. The impact of the war altered the demography of the village, leaving a large number of widows among the residents. After the end of the war in the eastern part of the country, the village became the target of mass postwar physical infrastructure development. We had previously researched this village in the period 2009–2011 as part of a project on youths and postwar recovery (Azmi et al., 2013). However, even after almost a decade of postwar physical infrastructure development, our field visits in 2018 revealed that the economic situation of the village had not improved.

Ponnamma and her husband lived in the village north of Batticaloa town. All of their three children had died under tragic circumstances. Ponnamma and her husband (age 72) cared for their three grandchildren. They all lived in a simple hut that had been provided to them by an NGO after the 2004 tsunami. Ponnamma was born in a nearby village, and she had fished in the river since age 10. Traditionally, women have not engaged in net fishing, but during the war, the *iyakkam* (Tamil Tigers) taught them the technique because the men were involved in fighting the war. Every evening (i.e. seven days per week), Ponnamma went to the river (located 1 km from her house) to cast her fishing net. She then returned home to prepare meals while her husband cared for the grandchildren. He took the grandchildren to school, helped them with their schoolwork, and fed them. Around 10 p.m., Ponnamma returned to the river to check the net, and she took any fish that had been caught to her brother, to sell.

Before receiving their house from the NGO, Ponnamma's family had endured several hardships: the loss of their children, wartime harassment, and four episodes of displacement after the tsunami in 2004. Ponnamma was worried about the future of her grandchildren: 'When I sit on the riverbank at night, all these worries come to my mind. Besides, we do not have a good market to sell our fish.' According to her, the lack of a market was a major problem for her and other female fishers because it meant they had to sell their fish for a low price. Instead of fishing, young men preferred to migrate for work in the Middle East, and some young women worked in a nearby factory.

Like Ponnamma, Gowry was a fisher. Through the tsunami housing scheme, she had received a house that was approximately 2 km from her fishing site. She explained that after the Tamil Tigers came to her village, most of the women both gleaned and used nets to catch prawns. Fishing with nets was not considered culturally appropriate for women, but Gowry had maintained the practice after the war, despite the hardship involved:

> Women are involved in making nets, repairing nets and tying the nets to catch prawns and fish. They also do domestic work and agriculture. Some weave Palmyrah [palm] baskets. Some women migrate to Puttalam [on the west coast] to work on the onion farms as labourers. They go as a family during the harvesting season. Young girls are not coming to this job [fishing]. They do not want to wet their feet in the salt water. Most of them are going to work in the garment factories in Trinco and Kanthalai.

In response to our question about how she had been affected by the tsunami reconstruction programmes, Gowry said,

> Many people did not get boats. . . . We were not given government assistance because this area was under the control of *iyakkam* [Tamil Tigers]. . . . However, big boat owners now take the labourers if they go far or they call them when they have work. This work is poorly paid and it is irregular.

In general, the women claimed that men had suffered deeply in the postwar situation. Gowry said, line shift 'Many men became weak after the war. They are having mental problems, they drink, they have heart problems, health problems related to high blood pressure. . . . They do not have work here in the village.'

Ponnamma and Gowry's stories revealed that both the tsunami and the war had severely affected their social well-being and ability to effectively fish, even though they had shifted from gleaning oysters to using nets to catch fish, which was a much more effective technique. It appears that fishing is not an option for young people and that the traditional fishing culture is disappearing as a consequence. It is mainly because of the women's efforts that the local community has survived. Men are excluded from fishing and other work in the village, due to their health problems and lack of alternatives created by other social, economic, and political processes that affect fisheries and other activities.

## Summary and conclusions

In our study, we found that distinguishing the different routes taken by women in Batticaloa District and Ampara District to become heads of households was essential for us to capture their different levels of vulnerability. We found a spectrum of female-headed households, including those with widows, sons, inactive husbands, and migrants.

The stories presented in this chapter reveal that the mobility and migration of the FHHs were intertwined with exclusion at individual and collective levels and that the moves that they and their family members made were gendered. Among the seven FHHs, the widows were the most vulnerable, especially in situations where they were the sole providers. They experienced an inferior situation and were struggling to make their lives better. FHHs who had male partners and either sons or a husband working abroad were in a better position due to their social identity and economic strength. The situation of FHHs was especially better when their male partners contributed economically. Our findings confirm that when women are the sole providers (in situations where husbands are not economically active or in situations where women have been abandoned) and have many dependents, their vulnerability is likely to be high. When male partners are absent, FHHs lack access to markets and are socially excluded.

Furthermore, we met women and men who, as a consequence of the civil war and the 2004 tsunami, were displaced far away from the fishing grounds. Others struggled with the available facilities in war-affected areas. The resettlement of people in places that were remote from their place of origin distanced them from their relatives and social networks in their ancestral villages. As a consequence, they were excluded from their traditional culture and social relations.

Fishing in the new settlement, in particular, was not regarded as a viable way of life. Postwar losses and the perceived higher incomes from migration are changing people's livelihoods. It is important to consider that the plans to develop and promote deep-sea fishing might put more pressure on small-scale fishers, particularly on women in fishing. As fishers move in and out of fishing in different stages of their life, they may find themselves temporarily excluded from a life in fishing.

Migration is considered a better way of life, but not a permanent one. Household members may migrate to foreign countries for work several times during their lifetime. Furthermore, all of the households we visited had one or more household members working abroad. Ironically, villagers who travel to faraway places for work may become integrated into the global economy, yet at the same time they may be excluded from social integration locally, particularly when their mobility is restricted due to poor infrastructure and means of travel. Overseas migration is an option mainly for young people, enabling some, mainly men, to find a well-paid job abroad and to save money. By contrast, women tend to benefit less from migration for work. Just after the civil war and the 2004 tsunami, the majority of migrants were women, but today the majority are young men. Although migration has become the main alternative to fishing, it is not necessarily a long-term solution to household's problems. Women who have returned after several years abroad face a difficult situation. Consequently, upon their return, life in their home village continues to exacerbate vulnerability and even produce precarity.

Each time fishers move to the Middle East for work outside the fishing industry, they experience both inclusion and exclusion in their home community and in their new community abroad. As a consequence of their migration, they face relational exclusion. In the aftermath of war, most of those who had been displaced to the new settlement remained there, whereas those who lived in the post-tsunami situation could not return to their old village due to land grabbing and development projects, all of which produced precarity. It follows that being included in fishing is not necessarily the best way to create a new living for them. At the same time, migration is not the only solution to overcome a precarious situation, given that most migrants in our research areas were unable to save large sums and send money home. Thus, the population we studied remained a disadvantaged group.

## Notes

1 Fazeeha Azmi and Ragnhild Lund, unpublished research on female fishers' vulnerabilities and well-being in Batticaloa.
2 The names of the villages and study participants are pseudonyms.

## References

Aasland, A., and Fløtten, T. (2001). Ethnicity and social exclusion in Estonia and Latvia. *Europe-Asia Studies*, *53*(7), 1023–1049.

Attanapola, C.T., Brun, C., and Lund, R. (2013). Working gender after crisis: Partnerships and disconnections in Sri Lanka after the Indian Ocean tsunami. *Gender, Place and Culture*, *20*(1), 70–86.

Azmi, F., Brun, C., and Lund, R. (2013). Young people's everyday politics in post-conflict Sri Lanka. *Space & Polity*, *17*(1), 106–122.

Azmi, F., Lund, R., Rao, N., and Manimohan, R. (2020). *Well-being and mobility of female-heads of households in a fishing village in South India. Gender Place and Culture.* https://doi.org/10.1080/0966369X.2020.1739003

Beegle, K., and van de Walle, D. (2019). *What can female headships tell us about women's well being? Probably not much* [Web log post]. Retrieved from https://blogs.worldbank.org/impactevaluations/what-can-female-headship-tell-us-about-womens-well-being-probably-not-much accessed 19/12/2019.

Béné, C., and Friend, R.M. (2009). Water, poverty and inland fisheries: Lessons from Africa and Asia. *Water International*, *34*(1), 47–61.

Benington, J. (2001). *Local partnership and social exclusion in the European Union: New forms of local social governance?* Retrieved from https://doi.org/10.4324/9781315011462 accessed 19/12/2019.
Blaikie, P., and Lund, R. (2010). *The Tsunami of 2004 in Sri Lanka: Impacts and policy in the shadow of civil war*. London: Routledge.
Buvinić, M., and Gupta, G.R. (1997). Female-headed households and female-maintained families: Are they worth targeting to reduce poverty in developing countries? *Economic Development and Cultural Change*, *45*(2), 259–280.
Cameron, A. (2006). Geographies of welfare and exclusion: Social inclusion and exception. *Progress in Human Geography*, *30*(3), 396–404.
Chant, S. (2004). Dangerous equations? How female-headed households became the poorest of the poor: Causes, consequences and cautions. *IDS Bulletin*, *35*(4), 19–26.
Coulthard, S. (2008). Adapting to environmental change in artisanal fisheries: Insights from a South Indian lagoon. *Global Environmental Change*, *18*(3), 479–489. Retrieved from https://doi.org/10.1016/j.gloenvcha.2008.04.003 accessed 19/12/2019.
de Haan, A. (1998). 'Social Exclusion': An alternative concept for the study of deprivation research. *IDS Bulletin*, *29*(1), 10–19. Retrieved from https://doi.org/10.1111/j.1759-5436.1998.mp29001002.x accessed 19/12/2019.
De Silva, D.A.M., and Yamao, M. (2007). Effects of the tsunami on fisheries and coastal livelihood: A case study of tsunami-ravaged southern Sri Lanka. *Disasters*, *31*(4), 386–404. Retrieved from https://doi.org/10.1111/j.1467-7717.2007.01015.x accessed 19/12/2019.
Department of Census and Statistics. (2017). *Poverty indicators: Department of census and statistics: Household income and expenditure survey – 2016*. Retrieved from www.statistics.gov.lk/poverty/Poverty%20Indicators_2016.pdf accessed 19/12/2019.
Department of Census and Statistics. (2018). *Household income and expenditure survey 2016: HIES final report 2016*. Retrieved from www.statistics.gov.lk/HIES/HIES2016/HIES2016_FinalReport.pdf accessed 19/12/2019.
Francis, P. (2002). Social capital, civil society and social exclusion. In U. Kothari and M. Minogue (Eds.), *Development theory and practice: Critical perspectives* (pp. 71–91). Basingstoke: Palgrave Macmillan.
Hall, D., Hirsch, P., and Murray Li, T. (2011). *Powers of exclusion: Land dilemmas in Southeast Asia*. Singapore: NUS Press.
INCLUSO. (n.d.). *The INCLUSO Manual: 1. Introduction*. Retrieved from www.incluso.org/manual/introduction accessed 19/12/2019.
Lehman, J.S. (2013). Relating to the sea: Enlivening the ocean as an actor in Eastern Sri Lanka. *Environment and Planning D: Society and Space*, *31*(3), 485–501. Retrieved from https://doi.org/10.1068/d24010
Ministry of Fisheries and Aquatic Resources Development & Rural Economy. (2018). *Fisheries statistics 2017*. Retrieved from www.fisheriesdept.gov.lk/web/images/pdf/Fisheries_Statistics_2018.pdf accessed 19/12/2019.
O'Reilly, D. (2005). Social inclusion: A philosophical anthropology. *Politics*, *25*(2), 80–88.
Rawal, N. (2008). Social inclusion and exclusion: A review. *Dhaulagiri Journal of Sociology and Anthropology*, 2, 161–80. Retrieved from https://doi.org/10.3126/dsaj.v2i0.1362 accessed 19/12/2019.

Ruwanpura, K.N., and Humphries, J. (2004). Mundane heroines: Conflict, ethnicity, gender, and female headship in eastern Sri Lanka. *Feminist Economics*, *10*(2), 173–205.

Sen, A. (2000). *Social exclusion: Concept, application, and scrutiny*. Manila: Asian Development Bank.

Sen, A. (2004). Elements of a theory of human rights. *Philosophy and Public Affairs*, *32*(4), 315–356.

Silver, H. (1994). social exclusion and social solidarity: Three paradigms. *International Labour Review*, *133*, 531–578.

Silver, H. (2007). *Social exclusion: Comparative analysis of Europe and Middle East youth*. The Middle East Youth Initiative Working Paper No. 1. Retrieved from https://papers.ssrn.com/sol3/papers.cfm?abstract_id=1087432 accessed 19/12/2019.

# 6 To migrate or not

## Social well-being and gendered household decision-making in fishing communities on the west and east coasts of Sri Lanka

*Nireka Weeratunge, Ramani Gunatilaka, Nadine Vanniasinkam, Mohamed Faslan, Dilanthi Koralagama and Nirmi Vitarana*

### Introduction

Fishing communities in Sri Lanka have long practised coast-to-coast migration because seasonal monsoons constrain fishing along the west and east coasts during parts of the year. Much of the migration takes place from the west coast to the northeast and east coasts during the six-month period of the southwest monsoon in Asia. In the past, entire households of small-scale fishers migrated (i.e. including women and children). However, with increased concerns about schooling and childcare, more women and children are now left behind at home. Other types of migration, both shorter in distance and more localized, most often involve male fishers who catch specific species of fish from waters farther along the same coasts on which their home villages are located. Relatively small numbers of men from east coast villages in-migrate to west coast villages for beach seine work. Additionally, both male and female members of fishing communities migrate either to engage in other livelihoods during the off-season for fishing or to seek employment overseas. Migration for work overseas has been practised only since the late 1980s.

This chapter focuses on seasonal coast-to-coast fisheries migration within Sri Lanka. The migration has been negatively affected by almost three decades of civil war (1983–2009), with both economic and social consequences. With an increase in both local and migrant fishers on the northeast and east coasts since the end of the war in 2009, contestations over the fish resource have increased, and the potential for these to escalate into ethnicized and/or regionalized conflicts has risen too, as elaborated in Chapter 4 in this volume. This chapter argues that the process of migration as a livelihood strategy is best understood in terms of micro-level and/or meso-level theoretical approaches to household decision-making and social networks. We posit that decision-making by individuals and households is motivated by underlying material, relational, and subjective factors that are encompassed by the pursuit of well-being rather than by poverty or resource scarcity. These factors are multifaceted and complex, and they differ across communities, gender, ethnicity, and/or religion and in terms of individual perceptions and aspirations. In the next section, we discuss the conceptual approach and methods. This is followed by a comparison of the context of fisheries in the sending district (Puttalam) on the west coast and the receiving district (Trincomalee) on the east coast and by an outline of the livelihood system and social relations. Thereafter, we present the main analysis of migration and household decision-making in four study villages within the two districts, followed by our conclusions.

## Conceptual approach and methods

Small-scale fishers' migration has been regarded as an adaptive livelihood strategy used by poor, resource-dependent people who move from areas of high poverty, high dependency on fishing, and depleted fisheries resources to areas of lower poverty, less dependency on fishing, and more-productive fisheries (Cripps, 2009; Jorion, 1988). This Malthusian approach has been challenged by Kraan (2009), Marquette et al. (2002), and Overå (2001), who propose that a combination of ecological, socioeconomic, and institutional factors motivates fishers to migrate. Overå (2001) argues that migration facilitates resource usage in new ecological niches, enabling small-scale fishers to increase their harvesting of fish in response to seasonal and long-term 'booms' in the ecological system, connected with the upwelling of the sea and the mobility of fish species. The role of translocal social and/or institutional networks in initiating and sustaining fishers' migration to expand their social and livelihood spaces has been emphasized (Kraan, 2009; Overå, 2001). Our approach follows a similar strand of perspectives, which have focused on individual and household decision-making (Stark, 1991), the role of social networks and institutions (Guilmoto and Sandron, 2001; Massey, 1990), and cultures of migration (Cohen and Sirkeci, 2011). Additionally, consideration has been given to gender approaches (Gamburd, 2005; Kabeer, 2007; Rao, 2009), especially to the different costs and benefits of migration to women and men and how they might experience this process differently.

We build on the aforementioned perspectives, focusing on what motivates members of fishing households to migrate in a social and/or institutional context. Our approach is based on a social well-being lens (McGregor, 2008; White, 2008), which challenges the primacy given to material well-being in development research and interventions by emphasizing relational and subjective dimensions, especially human agency. As elaborated in Chapter 1 and discussed in Chapter 4 in this volume, the analysis is framed by the three dimensions of material, relational, and subjective well-being. Social well-being is used here to understand both motivations and outcomes of livelihood strategies (including migration) in fishing communities. Coulthard (2012) emphasizes the significance of agency in decision-making and negotiating trade-offs in relation to well-being. A key assumption is that individuals are engaged in the pursuit of well-being through their livelihood choices, and although they are circumscribed within an institutional context, they are not merely victims of a larger system.

This chapter is based on an analysis that combined quantitative and qualitative methods, as outlined in Chapter 1 in this volume. A quantitative household survey (800 households) – the Survey on Migration and Livelihoods in Fishing Communities – was conducted by the International Centre for Ethnic Studies (ICES) in 2017 to address the lack of national data sets on livelihoods and incomes in fishing communities, seasonal fisher migrations, and social networks. A range of qualitative methods (136 individual interviews; 14 focus group discussions [FGDs]) was used to understand the gendered patterns and experiences of migration, as well as household decision-making. Female-headed households and those with left-behind spouses, not captured by the quantitative survey, were included in the qualitative interviews. In addition, 10 migrants from five households were traced from sending communities to receiving communities.

Four study villages[1] – Kadalpalli[2] and Cattiyur in Puttalam District and Selippur and Uppukadal in Trincomalee District – were selected for their high proportions of fishing households, extent of internal migration, ethnicity, size and adequacy for quantitative sampling, and links between sending and receiving communities. Cattiyur and Uppukadal are inhabited predominantly by Tamil Hindus, Kadalpalli predominantly by Sinhalese Catholics, and Selippur predominantly by Muslims. Fishers from Kadalpalli and Cattiyur are among the in-migrants to Selippur and Uppukadal.

## Context of fisheries in the sending and receiving districts

Sri Lanka's fish catch declined rapidly following the tsunami of 2004 to below prewar levels of 1983,[3] but thereafter, it increased until it peaked in 2014 and has since declined (Ministry of Fisheries and Aquatic Resources Development, n.d.). This trend is also observed in the two study districts, Puttalam and Trincomalee,[4] where production peaked in 2015 and 2012 respectively (Figure 6.1). The most recent assessment of marine resources off the coastal waters of the island revealed a sharp reduction to one-fifth of the fish resources available in the period 1978–1980.[5]

Fishing households in Trincomalee District have doubled since the 2004 tsunami, while those in Puttalam District have declined slightly since 2015 (Figure 6.2), generally corresponding with the increase in total fish catches between 2005 and 2015 (Ministry of Fisheries and Aquatic Resources Development, n.d.; National Aquatic Resources Research and Development Agency, n.d.).

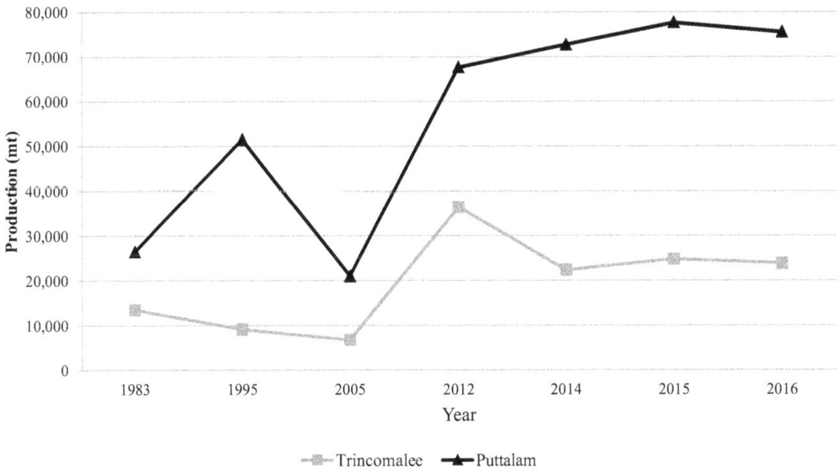


*Figure 6.1* Marine fish production in Puttalam District and Trincomalee District, Sri Lanka, 1983–2016

Source: Data from MFARD (2015) Annual Report: Fisheries Statistics 2015; MFARD (2018) Fisheries Statistics 2017

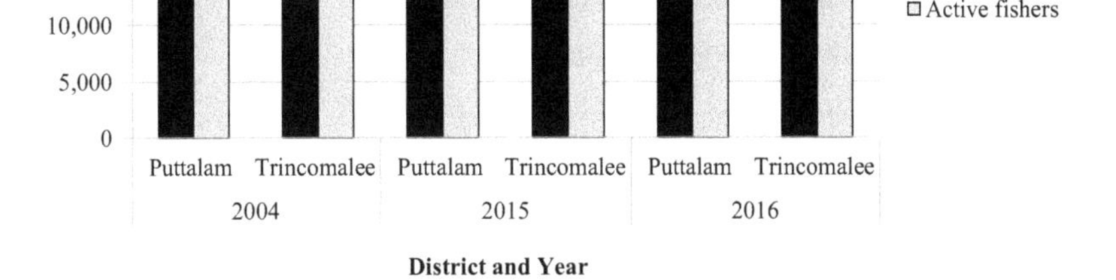


*Figure 6.2* Fishing households in Puttalam District and Trincomalee District, Sri Lanka, 2004–2016

Source: Data from NARA (2006) Sri Lanka Fisheries Yearbook 2003–2005; MFARD (2018) Fisheries Statistics 2017

The size of the fishing fleets, more than half of which comprise small motorized fibre-reinforced plastic boats,[6] has grown in both districts, but between 2005 and 2016, the growth rate was considerably higher in Trincomalee (3.7 times) than in Puttalam (1.7 times) (see Endnote 3). However, in 2016, Puttalam District had 9376 fishing vessels, almost twice as many as Trincomalee District had (Ministry of Fisheries and Aquatic Resources Development, n.d.).

In 2016, Puttalam, which is the largest fish-producing district in the country, produced three times more fish, with a fishing fleet almost twice the size of that in Trincomalee District and with 30% fewer fishers than Trincomalee (Ministry of Fisheries and Aquatic Resources Development, n.d.). However, the recent assessment of marine resources (see Endnote 3) indicates a higher rate of the depletion of fish off the northwestern coast than off the eastern coast. This probably explains why fishers from Puttalam are migrating from a technology-rich but more overfished area with a smaller fishing population to a technology-poor but less overfished area with a larger fishing population. Thus, it appears that neither overpopulation nor fisheries resource scarcity is the primary cause of seasonal migration between the two coasts.

Moreover, the empirical evidence does not suggest consumption poverty as a cause of internal seasonal migration. The results of an econometric analysis using per capita household consumption data confirmed a larger 'middle class' in the west coast study villages and a larger group of poor in the east coast study villages. In addition, the proportion of both male and female migrants from better-off households that owned a motorized boat was higher than from households with fishing crew or beach seine workers, indicating that relatively more migrants came from better-off households with fisheries-related assets. This point is elaborated in the following discussion on migration patterns. Overall, the evidence from the fisheries context in Sri Lanka contradicts the Malthusian theory of migration.

## Livelihood system and social relations

Fishing was the main economic activity and primary source of income for most households in all four study villages (Figure 6.3). However, households in the west coast villages were more dependent on fishing and fishing-related activities than were households in the east coast villages, which derived their incomes from other forms of manufacturing and services and from agriculture.

Fisheries assets differed in the four villages, with larger numbers of motorized boats in Kadalpalli on the west coast and Selippur on the east coast, compared with the other two villages, which had larger proportions of traditional vessels (canoes and rafts).[7] Cattiyur, on the west coast, had many more households engaged in beach seine operations than in the other three villages.

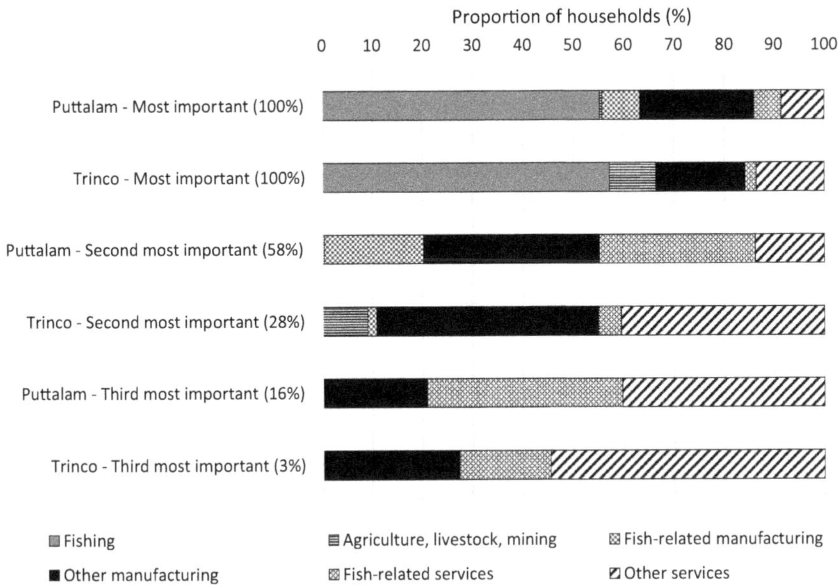


*Figure 6.3* Economic sector by most important source of household income in study villages, Puttalam District and Trincomalee District, Sri Lanka

Source: ICES Survey on Migration and Livelihoods in Fishing Communities 2017

With regard to the gender division of labour (Figure 6.4), men engaged in most fishing activities, whereas gleaning in the lagoons was the only fishing activity in which women participated, and the practice was largely confined to the two Tamil Hindu villages. In the west coast villages, women's engagement focused on post-harvest activities, such as disentangling fish from nets, sorting and processing fish, and the micro-trading or small-scale trading of fresh and dried fish. In the east coast villages, fewer women engaged in both fisheries-related activities and non-fisheries-related activities. On both coasts, men also engaged in post-harvest activities, such as dried-fish processing and trading both fresh fish and dried fish, but on a larger scale compared with the women.

In all four study villages, housework and childcare were mainly the domain of the women. The women also played a predominant role in managing household budgets, because they were the primary keepers of the earnings of both husbands and wives. Women were entrusted with the role of managing savings, especially in the form of jewellery, which, when pawned, provided a significant proportion of the financial investment required to purchase or replace fisheries assets. Apart from decision-making in fishing, which was the men's domain, and decision-making in daily household expenses, which was the women's domain, other decision-making within households, such as that relating to selling and processing fish, the purchase of fishing assets

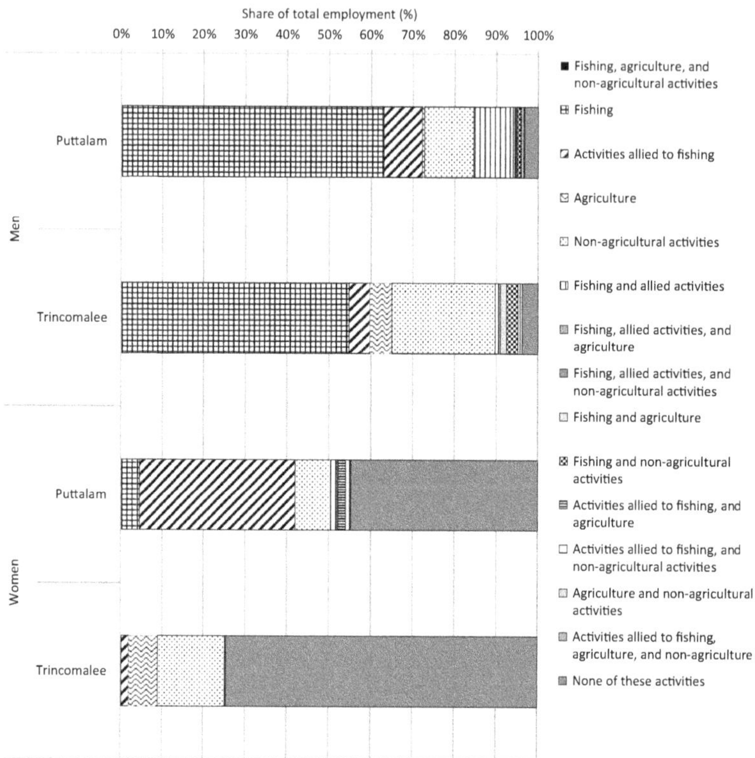


*Figure 6.4* Engagement of men and women by employment activity in study villages, Puttalam District and Trincomalee District, Sri Lanka

Source: ICES Survey on Migration and Livelihoods in Fishing Communities 2017

and household assets, and education, was mainly undertaken jointly, with varying degrees of influence by the spouses, depending on the household.

The social networks of the households (as elaborated in Chapter 10, this volume) were centred on family, kin, and neighbours. They were especially important for women, as the predominant form of marriage in three of the villages was matrilocal. Community-based organizations such as fisheries societies were important for men, while networks centred on religious institutions (Hindu temples, churches, and mosques) were important for both women and men. With the exception of female traders, men had more linkages to social networks outside their villages, especially those outside the district in which they lived, including wider political and institutional linkages to markets and the state. The external linkages of men and leaders of fisheries societies were higher in the west coast villages than in the east coast villages. Furthermore, as relatively fewer women engaged in fish processing and trading in the east coast villages, they had fewer linkages to networks outside their villages. Fishers who were mobilized through fisheries societies engaged in collective action relating to fisheries issues, such as the demand for subsidies from the government of Sri Lanka, and they organized for migration in the west coast villages and against migration in the east coast villages. Qualitative data from FGDs and interviews showed a greater emphasis on notions of relational well-being in west coast villages. This was supported by quantitative data from the household survey, which revealed that perceptions of livelihood support indicated higher reliance on networks both inside and outside the village in west coast villages. Thus, the emphasis placed on relational well-being in the west coast villages seemed to translate into stronger social linkages of support to pursue fishing, other livelihoods, and seasonal migration compared with in the east coast villages.

## Seasonal migration, social well-being, and household decisions

### *Migrations patterns*

Colonial administration records show that the seasonal migration of fishers and their family members from the west coast to the northeast and east coasts has been taking place for more than 150 years.[8] During both individual interviews and FGDs, members of fishing households in the two study villages on the west coast recalled up to two generations of seasonal migrants in their families.

There was little internal migration for fishing from Selippur, where households relied on a combination of marine fishing and lagoon fishing and/or farming to offset seasonal declines in access to resources and did not see the need for internal fisheries-related migration. Although the migration pattern for male fishers was similar in both villages on the west coast, the pattern for the women who accompanied them was different, with almost 40% of women from fishing households migrating seasonally from Kadalpalli, in contrast to approximately 10% from Cattiyur (Table 6.1). No women accompanied men from the two east coast villages.

The econometric analysis did not reveal a statistically significant correlation between household consumption and seasonal migration, but there appears to be a noteworthy difference in migration patterns according to the ownership of fishing vessels. Higher percentages of both men and women from households that owned a motorized boat migrated to fish compared with households that did not have a boat (Table 6.2). However, the household survey data revealed that of all male migrants, a slightly larger share (45%) came from those without a fishing vessel (i.e. crew and beach seine workers), followed closely by those who own a motorized boat (41%), whereas 66% of all female migrants came from households that owned a motorized boat (percentages not shown in Table 6.2). Women from boat-owning households accompanied their husbands to support their fishing activities, to cook and to maintain the temporary shelter for both their husbands and the accompanying crew members, and to engage in their own

*Table 6.1* Migrants, who migrated for fishing and related livelihoods, from fishing households in study villages, Puttalam District and Trincomalee District, Sri Lanka, 2017

| *GN division** | *Men* | | *Women* | | *Migrant HHs*** | | *All fishing HHs* | |
|---|---|---|---|---|---|---|---|---|
| | *No.* | *Fishing HHs (%)* | *No.* | *Fishing HHs (%)* | *No.* | *Fishing HHs (%)* | *Total No.* | *Total %* |
| West Coast (Puttalam District) | | | | | | | | |
| Kadalpalli | 145 | 76.3 | 71 | 37.4 | 146 | 76.8 | 190 | 100 |
| Cattiyur | 113 | 72.4 | 16 | 10.3 | 115 | 73.7 | 156 | 100 |
| East Coast (Trincomalee District) | | | | | | | | |
| Uppukadal | 23 | 19.0 | 0 | 0 | 23 | 19.0 | 121 | 100 |
| Selippur | 6 | 4.2 | 0 | 0 | 6 | 4.2 | 142 | 100 |
| Totals | 287 | 47.1 | 87 | 14.3 | 290 | 47.6 | 609 | 100 |

Note: * GN Division – Grama Niladhari Division; **HHs – households

Source: ICES Survey on Migration and Livelihoods in Fishing Communities, conducted in 2017

*Table 6.2* Migration of male fishers and accompanying female family members, by ownership of fishing vessel, from study villages, Puttalam District, Sri Lanka, 2017

| *Vessel type* | *Migrant male fishers* | | *Accompanying female family members* | | *All fishing HHs** |
|---|---|---|---|---|---|
| | *No.* | *%* | *No.* | *%* | *No.* |
| Motorized fibre-reinforced plastic boat** | 107 | 89.9 | 55 | 46.2 | 119 |
| Traditional vessel (canoe or raft)*** | 33 | 55.9 | 12 | 20.3 | 59 |
| Beach seine owner | 2 | 40.0 | 1 | 20.0 | 5 |
| No fishing vessel (crew/worker) | 117 | 72.7 | 15 | 9.3 | 161 |
| Totals | 261 | 75.4 | 83 | 24.0 | 346 |

Notes: * Households; ** in Sri Lankan fisheries reports and statistics, the abbreviation OFRP (outboard-engine fibreglass reinforced plastic) boat is commonly used to refer to small motorized fibre-reinforced plastic boats; *** in Sri Lankan fisheries reports and statistics, the abbreviation NTRB (non-motorized traditional boat) is used to refer to traditional canoes and rafts

Source: ICES Survey on Migration and Livelihoods in Fishing Communities, conducted in 2017

fish-processing activities. Wives of crew and beach seine workers were generally not provided with accommodation in migrant workers' camps and therefore were less likely to migrate.

Seasonal migration destinations have changed historically in the two west coast villages. Before the ethnic civil war in 1983, the main destination of fisher households from both villages was the northeast coast in Mullaitivu District (Figure 6.5), although some households migrated to locations along

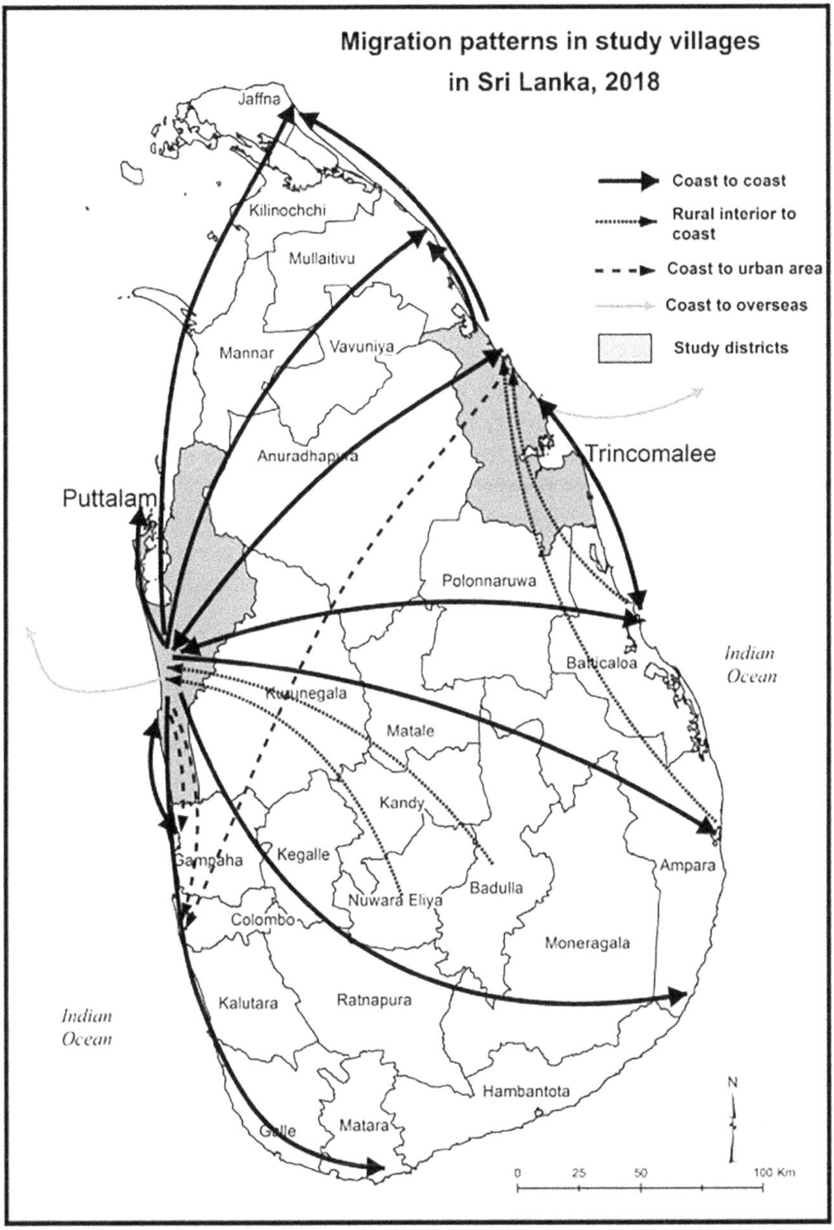


*Figure 6.5* Migration patterns in study villages, Puttalam District and Trincomalee District, Sri Lanka

Source: ICES/NTNU data generated in 2018 by the project Migration and Collectives/Networks as Pathways out of Poverty: Gendered Vulnerabilities and Capabilities amongst Poor Fishing Communities in Asia, superimposed on a base map prepared by the Survey Department of Sri Lanka, Colombo, 2016

the entire east coast, from Trincomalee District to Ampara District. During the three decades of the war (1983–2009), seasonal migration reduced overall, but most west coast fishers migrated to locations in Trincomalee District, due to the presence of the navy and some measure of protection that it offered. Some fisher households started to migrate to Baththalangunduwa, a fishing village on an island off the west coast (Puttalam District), to the north of their villages. Since the war ended in 2009, fishers and household members have continued to migrate to the east coast and to return to prewar locations on the northeast coast.

### *Reasons for migration*

In the studied communities, we assessed the households' reasons for migration, enabling and disabling factors, and benefits and costs as embedded in their decision-making processes. On both the west and east coasts, people perceived the main reasons for internal migration as material and relational. In the west coast villages, people explained out-migration by indicating their various needs, which were to continue fishing during the off-season for fishing when the southwest monsoon occurs in Asia, to avert risks to life by avoiding fishing in rough seas, to pay off debts and set aside savings, and to continue a tradition followed by their grandparents:

> During the *warakang* (off-season), we cannot fish in our sea. The sea is so harsh. We cannot earn much due to the poor catches. The estuary is so dangerous. So, we are afraid to do fishing here, but we need to do our job. We need to eat and meet our expenses. If we borrowed money from a *seetu* (rotating credit association) we would need to pay it back. So, we need to migrate. In the sea in Selippur, we catch more fish. The prices do not fluctuate from time to time as they do here. We can get a fixed price throughout the day.
>
> (Lalani, wife of a fisher, Kadalpalli, Puttalam)[9]

> For us here, migration is very historical. However, the people in that area [host location] now do not know about our migration pattern. Before the war started, we used to work in Mullaitivu [on the northeast coast]. When the intensity of the war escalated, we left all our belongings, boats, engines, nets, and all, and came back.
>
> (Participant in men's FGD, Kadalpalli, Puttalam)

The men in the west coast villages also referred to returns from pursuing large, high-value target species, in the wake of the seasonal migration of flying fish:

> There are the flying fish, which migrate [along the northeast coast]. Following the flying fish are tuna and Spanish mackerel, which feed on

> them. We had knowledge about these migrating schools of fish for a long time. They are only there for three months, from April to July. So that is why we try to get there before April. . . . Many of the fish breeding grounds are there. Tuna, for example, hatch on the east coast, and migrate north before coming to our west coast here.
>
> (Participant in men's FGD, Kadalpalli, Puttalam)

Fishers in the east coast villages perceived the underlying reason for in-migration as the easy access to east coast resources, due to a sea-is-for-all fisheries policy in Sri Lanka:

> All the fishermen from Mullaitivu to Pudavaikattu [fishing village] fish in the sea in this area. So, when 300 new boats [of migrants] come, there is not enough for everyone, but the fisheries authorities say that the sea is for all and there are no restrictions. They don't think of personal losses. There are no regulations for building *wadi*. . . . We suffer great losses due to migration.
>
> (Mohamed Usuf, male, beach seine owner, Selippur, Trincomalee)

However, traders and/or boat owners from the east coast who sponsored in-migrants cited a shortage of labour to harvest the abundant fish resource during the season as the main reason for in-migration.

### *Enabling and disabling factors*

The primary enabling factors in internal migration on both coasts were identified as relational. Women and men in the west coast villages indicated that in a prevailing culture of migration, out-migration was enabled by the strength of social networks in the village, mobilization in small groups, and the assurance of care for family members left behind. Additionally, links to fish traders and/or boat owners who acted as sponsors in the host villages were critical:

> We migrate with the help of a fish trader known to us for more than 10 to 15 years. He is like our relative. As a strategy, we transfer our boats to him. He pretends that we are his debtors. He says to the locals, 'I need to get back my money. So let them fish here. Then they can pay me their debt. Otherwise, how can I get my debts?'
>
> (Lalani, wife of migrant fisher, Kadalpalli, Puttalam)

The role of local traders in bringing in migrants was also emphasized by fishers in the east coast villages. In addition, men on both coasts referred to the support of fisheries authorities and politicians in the south for the migration process. Men in the west coast villages also indicated the importance of support from church authorities and the importance of material factors, such as their own superior skills in fishing relative to the skills of the host population, knowledge of the fishing grounds in the migration sites, and knowledge of the language spoken in the host villages:

> We go there to follow this migrating fish. We call them *le malu* ['blood fish']. It is a big harvest for our country. Those days [in the past], there were only a handful of boats to catch this fish there. It is people from here, our grandfathers, who first identified it and started to harvest it. The fish migrate at that time, so if we don't catch them, we lose this harvest. Those days, people over there were farming. They learned to fish from us. Now they also know of this fish resource.
>
> (Participant in men's FGD, Kadalpalli, Puttalam)

Men in the east coast villages cited other relational factors for in-migration, such as the fisheries authorities' partiality shown towards migrants and specific ethnic groups, the weakness of politicians in the north and east, and disunity in their own fisheries societies:

> Government officials are biased. They catch only the boats [using banned nets] of Muslims but not the boats of the Tamils and Sinhalese. . . . What is banned in Trincomalee is not banned in other districts. We do not have issues or conflict with outsiders [migrants], but government officials and media are biased, and they take the side of outsiders.
>
> (Participant in men's FGD, Selippur, Trincomalee)

On both coasts, disabling factors in internal migration were relational and material. Women and men in the west coast villages indicated that migration was constrained by the need for the care and education of children in their home villages in their absence, the increasing opposition to and restrictions placed on in-migration by host communities, and the access to alternative incomes from business, permanent employment, or remittances from family members abroad:

> Normally we migrate in March and stay until July or August in Selippur. Up to now, I would take my child [age 3] with me, but from next year onwards I will not be migrating because I need to take care of my child and her schooling. Mothers nowadays don't leave their children behind and migrate.
>
> (Dinesha, wife of migrant fisher, Kadalpalli, Puttalam)

> We couldn't migrate to Mullaitivu as before, due to restrictions, such as getting permission from their fisheries societies. There were no

> fisheries societies in Mullaitivu then. We have not been migrating there for four to five years. My wife and I decided not to migrate due these barriers.
>
> (Ramakrishnan, raft fisherman, Cattiyur, Puttalam)

Additionally, men in the west coast villages referred to increasing competition between migrant and local fishers, the use of illegal fishing gear and methods by local fishers, conflicts over fishing gear, and quotas placed by fisheries or local authorities on migrants' boats. Men in the east coast villages indicated overcrowding and a lack of landing spaces for boats on the beach, as well as unity among fisheries societies that had restricted migrant fishers from entering their villages on several occasions. In addition to their maternal role, women in the west coast villages indicated the inability of their husbands to manage earnings at the migration sites as disabling:

> When migrating out, he would accompany a *mudalali* [trader]. The *mudalali* would advance money and my husband would already advance the money to the worker even before going, unlike other people in the village, and a lot of money would be spent to maintain the worker. I could not go in the first year because I was pregnant. By the time I was able to accompany him in the third year, it was too late because he had got into too much debt and incurred losses by then.
>
> (Nirosha, left-behind wife of fisher, Kadalpalli, Puttalam)

### *Benefits and costs*

All study participants indicated that the primary benefits of internal migration were material. Women and men in the west coast villages cited the benefits as the considerably higher earnings from fishing on the northeast and east coasts, the redeeming of pawned jewellery, the ability to settle debts, and the cash injections into the local economy of the host villages. Women in the west coast villages similarly referred to higher earnings, specifically from dried-fish processing:

> We have been migrating for over 25 years now. At the migration site, we provide support to remove fish from the nets, cook and, especially, we have to treat the fish labourers well because they could leave if not treated well. We also do dried-fish processing at the *wadi* site. The host community villagers purchase dried fish from us. We can sell dried fish to anyone while we are there but the [fresh] fish we catch must only be sold to the trader who sponsors us to migrate.
>
> (Participant in women's FGD, Kadalpalli, Puttalam)

Men indicated several subjective benefits of internal migration, such as a change of environment and a sense of camaraderie while at the places to which they migrated:

> It is fun in the *wadi*. We bathe, play cards, which is our only entertainment, drink plain tea, and go to sleep. We can't go anywhere. We have a radio, but no television. There is no electricity in the *wadi*.
>
> (Participant in men's FGD, Cattiyur, Puttalam)

By contrast, men and women in the east coast villages did not perceive any benefits from in-migration, other than for the fish traders who sponsored migrants and therefore received bigger profits. On both coasts, the costs of internal migration were expressed in material, relational, and subjective terms. Women and men in the west coast villages indicated material costs, such as anticipated earnings not being realized and the loss of fisheries assets due to conflicts with the host communities, both of which resulting in them incurring further debts. The social and subjective costs identified by women and men in the west coast villages were vulnerability and/or security of women at the migration sites and nostalgia regarding their home. Women in the west coast villages emphasized relational costs, such as separation of family members and neglect of children's education, but also referred to subjective costs, such as loneliness at home or difficult living conditions in migrants' camps, and the material costs, such as the incurrence of debts when they failed to accompany their migrating husbands:

> I liked that my husband went there to the north for two seasons because we were able to build our house with the money he earned there. Now, I don't want him to migrate any longer because we have two children here and he needs to be here. When he is over there, I am lonely and every day I am thinking about when he will be coming back home. There is nobody here for me. I think about him going so far away. I have told my husband that with our house now completed we have enough, so 'Don't go anywhere. We are fine here as we are.'
>
> (Shanthi, female fish trader and wife of lagoon fisher, Kadalpalli, Puttalam)

Relational and subjective costs identified by men in the west coast villages were bad working conditions and the deterioration of health in migrant workers' camps, especially for beach seine workers. Men in the east coast villages perceived both material and relational costs, including the use of illegal fishing gear and methods by migrants, the loss of fisheries resources and incomes for local fishing households, conflicts between migrants and local people, the pollution of beaches, and the spread of disease and sex work:

> We used to catch tuna and mackerel. Now, if we catch these one day, the next day there is none. The fish resources have depleted because of the *naaivalai* [trammel nets] used by people from Kadalpalli.
>
> (Participant in men's FGD, Uppukadal, Trincomalee)

> We do not have enough space and fish in the sea due to migrant fishers. We do not have any jobs other than fishing.
>
> (Ruthrakumar, fisherman, Uppukadal, Trincomalee)

Most women in the east coast villages said that they did not have much interaction with migrant fishers, because migrant workers' camps were not located near their houses.

### *Household decisions*

The study participants revealed that decisions relating to migration were made jointly by husbands and wives in the west coast villages. Household members were unlikely to migrate if their household had a stable alternative income during the off-season for fishing or if their household received remittances from family members overseas. While wives of boat owners needed to decide whether or not to migrate, wives of crew or beach seine workers did not migrate, because they were not provided with accommodation in the migrant workers' camps. The specific migration location and the period of migration were decided by a combination of factors that were relational and material. When making their decisions, men considered past experience in the migration location, information from their peers about potential migration locations that would enable a good catch, the ownership of appropriate fishing gear, the offer made by the sponsoring fish trader or boat owner in terms of advances, fish prices (in the case of fishers who owned their boats), and wages (in the case of crew and beach seine workers):

> Villagers who migrate to those areas [northeast and/or the east] share their experience, good income, and stories with us. They [villagers] motivate us to decide on the next migration site. However, we found that Kalmunai was not good for us, due to the heavy winds, wave patterns, high cost, and not [being] suitable for our fishing method. Site selection is highly dependent on the fishing method we are employing. In fact, we came back to Nilaveli [coastal town, Trincomalee District] for the next two years.
>
> (Dinesh, migrant fisherman, Kadalpalli, Puttalam)

Women were influential in decisions about whether to accept the offer of the sponsoring fish trader. The decisions of wives of boat owners whether to accompany their spouse or not was mainly based on relational factors linked to the stage of the life cycle and composition of the household, including the willingness of the mother-in-law or mother to migrate instead. Women with grown children generally decided to migrate, while those with infants, toddlers, or school-age children increasingly weighed their options to migrate. Their decision not to migrate was influenced by gender norms concerning the role of motherhood, which in Kadalpalli was reinforced by the Church:

> Through the Church, we have given advice during Sunday mass about the need for at least the mother of the family to stay back with the children and look after the household while the father and other men migrate.
>
> (Father Lasantha, Catholic priest, Kadalpalli, Puttalam)

However, sometimes subjective factors in relation to the security of children left behind, especially of girls, contributed to the women's decision to migrate. While relational and subjective factors caused some women to remain at home, they also had to face the relational cost of a long separation from their spouse, the material costs of not being able to earn from dry processing fish, and the risk and uncertainty of whether their husband would manage his finances effectively:

> In families that have money, the women don't migrate with the husbands. They also don't migrate if the husband works in a beach seine camp. I have to migrate with my husband because we have our own boat. People tell me, 'You are leaving your only child and going.' I tell them that we visit him during the holidays and he also comes to us during the holidays.
>
> (Sivamalar, dried-fish processor and wife of migrant fisherman, Cattiyur, Puttalam)

Some male fishers, when deciding to migrate, acknowledged that the material benefits had relational and subjective costs, especially in terms of separation from children and relations involving affection and love:

> We keep our son at my mother in-law's house in Mullaitivu for schooling. . . . We get to work only if we migrate here [Selippur] but we lose our son's love. You lose one thing to gain something else. . . . If he has holidays, we go and pick him and bring him here.
>
> (Vigneshawaran, migrant fisherman, Cattiyur, Puttalam)

## Conclusions

The findings from our study revealed that the internal coast-to-coast migration by small-scale fishing communities from Puttalam District to Trincomalee District in Sri Lanka was motivated not by poverty in the sending districts but by seasonal opportunity in the receiving region, as also argued by Overå (2001) and Marquette et al. (2002) in the case of West Africa. Households' decisions on whether to migrate are better explained by the material, relational, and subjective factors that underlie the pursuit of well-being, as elaborated by McGregor (2008) and White (2008).

The reasons for migration were largely material and relational from the perspective of both migrant and host communities. The relational factors emerged as particularly strong for both enabling and disabling migration, especially the role of social networks inside and outside communities, as elaborated in Chapter 10 in this volume. Migrant households on the west coast perceived out-migration as entailing a combination of material, relational, and subjective benefits and costs. However, host households on the east coast perceived more costs than benefits from in-migration. On both coasts, women tended to bear more of the burden of relational costs. Thus, reconciling the pursuit of well-being between migrant and host communities and between women and men who incur differential benefits and costs remains a challenge.

## Notes

1 Each study village corresponds to a Grama Niladhari (GN) Division, the smallest administrative unit in Sri Lanka.
2 The village names used in this chapter are pseudonyms, in accordance with ethnographic convention.
3 Ministry of Fisheries and Aquatic Resources Development, unpublished annual report on fisheries statistics dated 2015.
4 Puttalam District comprised two fisheries districts for administrative purposes, Chilaw and Puttalam, whereas Trincomalee District remains one district for both administrative and fisheries purposes. For this chapter, the data for both fisheries districts were amalgamated in the case of Puttalam, to compare the sending and receiving districts, Puttalam and Trincomalee respectively.
5 Personal communication in 2019 from the National Aquatic Resources Research and Development Agency (NARA), on the preliminary findings of the Survey of Regional Resources and Ecosystem off Bay of Bengal: Sri Lanka.
6 In Sri Lankan fisheries reports and statistics, the abbreviation OFRP (outboard-engine fibreglass reinforced plastic) boat is commonly used to refer to small motorized fibre-reinforced plastic boats.
7 In Sri Lankan fisheries reports and statistics, the abbreviation NTRB (non-motorized traditional boat) is used to refer to traditional canoes and rafts.
8 R.L. Stirrat, 1972, notes on Ceylon Administration Reports, 1867–1887, held in personal archive material.
9 The names of all study participants are pseudonyms, in accordance with ethnographic convention.

## References

Cohen, J.H., and Sirkeci, I. (2011). *Cultures of migration: The global nature of contemporary mobility*. Austin, TX: University of Texas Press.

Coulthard, S. (2012). Can we be both resilient and well, and what choices do people have? Incorporating agency into the resilience debate from a fisheries perspective. *Ecology and Society*, *17*(1), 4. Retrieved from http://dx.doi.org/10.5751/ES-04483-170104 accessed 19/12/2019.

Cripps, G. (2009). *Understanding migration among traditional fishers of West Madagascar*. Blue Ventures Conservation Report for ReCoMaP. Retrieved from https://blueventures.org/wp-content/uploads/2015/10/Cripps_2009.pdf accessed 19/12/2019.

Gamburd, M.R. (2005). Lentils there, lentils here: Sri Lankan domestic labour in the Middle East. In S. Huang, B.S.A. Yeoh, and N.A. Rahman (Eds.), *Asian women as transnational domestic workers* (pp. 92–114). Singapore: Marshall Cavendish.

Guilmoto, C.Z., and Sandron, F. (2001). The internal dynamics of migration networks in developing countries. *Population: An English Selection*, *13*(2), 135–164.

Jorion, P. (1988). Going out or staying home: Seasonal movements and migration strategies among Xwla and Anlo-Ewe fishermen. *Maritime Anthropological Studies*, *1*(2), 129–155.

Kabeer, N. (2007). *'Footloose' female labour: Transnational migration, social protection and citizenship in the Asia region*. IDRC Working Papers on Women's Rights and Citizenship. Retrieved from www.idrc.ca/sites/default/files/sp/Documents%20EN/WRC-WP1-Kabeer-Migration.pdf accessed 19/12/2019.

Kraan, M.L. (2009). *Creating space for fishermen's livelihoods: Anlo-Ewe beach seine fishermen's negotiations for livelihood space within multiple governance structures in Ghana*. African Studies Collection 19. Leiden: African Studies Centre.

Marquette, C.M., Koranteng, K.A., Overå, R., and Aryeete, E.B. (2002). Small-scale fisheries, population dynamics, and resource use in Africa: The case of Moree, Ghana. *Ambio*, *31*(4), 324–336.

Massey, D.S. (1990). Social structure, household strategies, and the cumulative causation of migration. *Population Index*, *56*(1), 3–26.

McGregor, A. (2008). *Well-being, poverty and conflict*. ESRC Research Group on Well-being in Developing Countries, Briefing Paper 1/08. Retrieved from www.bath.ac.uk/soc-pol/welldev/research/bp/bp1-08.pdf accessed 19/12/2019.

Ministry of Fisheries and Aquatic Resources Development. (n.d.). *Fisheries statistics 2017 [2016]*. Retrieved from www.fisheries.gov.lk/web/images/downloads/pdfs/fisheries_statistics_2017.pdf accessed 19/12/2019.

National Aquatic Resources Research and Development Agency [NARA]. (n.d.). *Sri Lanka fisheries yearbook 2003–2005*. Retrieved from www.erepository.nara.ac.lk/bitstream/handle/1/1188/NR-SOCIO-19-FULL-TEXT.pdf?sequence=2&isAllowed=y accessed 19/12/2019.

Overå, R. (2001). *Institutions, mobility and resilience in Fante migratory fisheries in West Africa*. CMI Working Paper WP 2001:2. Bergen: Chr. Michelsen Institute.

Rao, N. (2009). *Gender differences in migration opportunities, educational choices and wellbeing outcomes*. Development Research Center on Migration, Globalisation and Poverty, University of Sussex. Retrieved from www.researchgate.net/publication/265746792_Gender_Differences_in_Migration_Opportunities_Educational_Choices_and_Wellbeing_Outcomes accessed 19/12/2019.

Stark, O. (1991). *The migration of labour*. Oxford: Blackwell.
White, S.C. (2008). *But what is wellbeing? A framework for analysis in social and development policy and practice*. Unpublished manuscript, University of Bradford. Retrieved from http://people.bath.ac.uk/ecsscw/But_what_is_Wellbeing.pdf accessed 19/12/2019.

# 7 Fishing in distant waters

## Issues of identity and well-being among migrant fishers on the west coast of Tamil Nadu

*Nitya Rao and C.M. Pratheepa*

### Introduction

Fishers' movements and migratory patterns follow the movements of fish, and in the process, they blur the temporal and spatial boundaries that are often referenced in migration research (Abobi and Alhassan, 2015). There is limited literature on migration and labour mobility in fishing communities in India and even less on the implications for gender relations and well-being outcomes. The existing literature focuses mainly on the movement of agricultural labourers who practise seasonal fishing during the lean agricultural season by taking on lower-paid and less-skilled work in the fishing industry (Bavinck, 2014; Swathilekshmi and Johnson, 2015). In this chapter, we seek to fill the knowledge gap by focusing on Kanyakumari District in Tamil Nadu, on the west coast of India, because the residents represent an exception to the aforementioned pattern. Fishers in the district display three distinct patterns of movement: multiday fishing; seasonal intrastate and/or interstate migration using harbours inside or outside the state; and migration overseas for work. In all three cases, the fishers not only seek to enhance their incomes and earning potential but also seek to uphold their identities as skilled fishers (Sathiadhas and Prathap, 2009).

The focus of this chapter is on multiday deep-sea fishing trips (ranging from eight to 45 days), which accounts for more than 80% of the fishers' movements. While migration drivers vary across the different streams of migration, decisions relating to multiday deep-sea fishing are shaped particularly by masculine fisher identities, especially as 'shark hunters'. Apart from earning money, which clearly is significant in their pursuit of well-being, the fishers – who mainly belong to the Mukkuvar caste and are Roman Catholic by faith – derive their satisfaction from displaying their skills in confronting and overcoming serious risks (Narendran, 2016; Pollnac and Poggie, 1988) (on Puttalam fishers' migration motivated by their fishing skills, see Chapter 6, this volume). In this chapter, we attempt to unpack the complexity of movement involved in multiday fishing as a dynamic process and its gendered implications by focusing specifically on the fishers' access to resources, their skill development, and their expressions of agency and identity.

Apart from taking place in a highly heterogeneous, uncertain, and competitive environment (Acheson, 1981), fishing is a highly risky enterprise because it carries risks to both life and livelihood. Nevertheless, fishers persist, seeing risk taking as the only way of dealing not only with a range of existing vulnerabilities but also with new precarities arising from the industrialization of fisheries (Rigg et al., 2016). However, their risk-taking behaviours, including being adventurous, courageous, and aggressive, are represented as central to their identity as fishers. Their social and cultural values celebrate traditional 'masculine' traits of bravery and fearlessness, the thrill of the hunt, the challenge of facing the power and expanse of the sea, and the adventure of pitting themselves against the elements (Binkley, 1995; Pollnac et al., 1998). Such constructions of masculinity provide an occupational identity for male fishers and reflect a distinct fishing way of life (McGoodwin, 2001); at the same time, they marginalize women's role in fishing enterprises.

In the studied context, the Church has been active not only in ensuring that women gain a higher education but also in two other critical ways: women are encouraged to be good wives, who actively participate in prayers for the safe return of the men in their households, and they are helped to play a supportive but critical role, through taking over complete responsibility for the management of the domestic domain, including raising the capital required for the fishing enterprises (Hapke and Ayyankeril, 2018; Ram, 1991). This is reflected in both the provision of training and skills for effective money management and the provision of capital through savings groups, which are run by neighbourhood groups (*anbiams*). In this chapter, we take a relational perspective on gender in order to unpack the gendered discourse of skills and agency, focusing specifically on the gendered divisions of work, roles and responsibilities, access to resources such as social institutions, and decision-making agency (Moore, 1994).

This chapter is based on a survey, conducted in 2017, of 103 fishing households in a cluster of villages in Thoothoor panchayat,[1] on the northernmost tip of Kanyakumari District, bordering Kerala. In addition to the survey, in-depth interviews were conducted with women and men in fishing households and with key informants, and focus group discussions were held. In the next section, we briefly recount the history of deep-sea fishing. Thereafter, we explore the interlinkages between migration, movement, and identity in constructing narratives of both risk and skills. Then, the implications for gendered identities and well-being are explored, followed by some brief conclusions.

## Deep-sea shark fishing in Thoothoor: the historical and social context

Before examining the specific notions of skills and identity of male fishers in the study context, we briefly discuss the cultural importance of the fish targeted by the male fishers – the shark – and the history and patterns of fishers' movements since the late 1950s, which have changed in response to conflicts with local fishers, intermediaries in markets, and shifts in technology (David, 2013).

Globally, sharks are more than just marine resources. They are regarded equally with fear and reverence, where many groups of fishers consider sharks as totems, symbolizing the might of the sea. The ancient Polynesians worshipped sharks as gods of the sea (Torrente et al., 2018), and their rituals shared similarities with the ancient Tamil ritual of worshipping sawfish teeth. Since sharks were feared, male fishers who caught sharks were lauded for their courage and valour, which elevated a livelihood act to a heroic feat (அகநானூறு [*Agananooru*], n.d.). The Tamil word *sura vettai* ('shark hunting') referred to catching sharks, and the practice was celebrated as the best type of hunting at sea. The nomenclature explains fishers' perceptions, distinguishing 'shark hunting' from other forms of fishing and justifying the movement of the male fishers to distant, deep waters to catch sharks. In this respect, an understanding of the cultural history of the Mukkuvars is important because the Mukkuvars are the dominant group in the study locality, and their name is derived from the element *mukku*, which means 'going underwater'.

The Mukkuvars are known for their excellent seafaring skills and fishing techniques (Hornell, 1908) and are mentioned in the Sangam literature[2] as traditional fisherfolk of the *neithal* (coastal) region or coastal landscape. In the Sangam literature, the male fishers are referred to as having *maram thozhil* (a hereditary occupation), not as practising *Thoondil velai* (hook-and-line work). Such men had the much envied opportunity of drawing in unusually large single fish and the occasions were the source of many tales and a certain amount of individualistic pride (Ram, 2008). Today, the male fishers regard themselves as the best fishers on the west coast. They are labelled 'brave men' (*veerakkarar*) and are much sought after by mechanized boat owners, even from distant regions. They take immense pride in their hook-and-line fishing, bravery, and skill (Ram, 2008). Their caste identity signifies a particular set of skills, which includes different forms of knowledge (social, technical, and practical) but equally also 'bodily disposition, technical dexterities and the capacity to innovate' (Sundar, 2018, p. 2).

Thoothoor panchayat was selected for testing the government of India's 'gill net package' in the 1960s. However, this did not lead to the mechanization of boats, because the local fishers were able to use the technology on their traditional small fishing vessels (*vallam*). As a few older men noted during a focus group discussion (FGD), even when they did not have high horse-power engines, they were able to use the wind, stars, and sun to navigate to the deep-sea areas inhabited by sharks. In the late 1980s, a full-fledged fleet of 32 mechanized boats was established at Thoothoor; the boats were 9.75–10.35 m in length, and used 50 m longlines for shark fishing. By 1992, within five years of starting fishing activities from harbours in Kerala and farther north in Mangalore, the boats had moved to bases in the west coast states of Maharashtra (Raigad District), Gujarat (Porbander city), and Goa, all of which enabled access to a wider stretch of the continental shelf (up to 200 m) compared with the west coast of Tamil Nadu or Kerala.[3]

In the early years, when fishers crossed state boundaries to fish, problems emerged as a result of the coast guard demanding proof of the fishers' identity and their registration documents. The problems led to the formation of the Association of Deep Sea Going Artisanal Fishermen (ADSGAF) in Thoothoor in 1992, which then negotiated the issue of identity cards with the state.[4] Following this recognition, the fleet expanded, and almost 600 boats started fishing farther north along the coast, which led to increased conflict with fishers in the host communities. In 1994, local small-scale fishers in Karwar city (in the state of Karnataka) burnt four boats, and tensions also heightened in other locations (V. Vivekanandan, personal communication, 4 February 2019).

By the mid 1990s, the Thoothoor fishers had to change their tactics. They increased their boat size to enable them to prolong their trips to the same fishing grounds rather than use local harbours as bases for their fishing activities. They also diversified from using only bottom longlines, which were more suitable for fishing on the continental shelf, to using surface longlines, which helped them to access oceanic shark species beyond the shelf (Kasim and Vivekanandan, 2011, p. 9). Thoothoor panchayat is renowned as the only cluster of fishing villages in India where fishers catch sharks in deep waters by using the hook-and-line method (Fernando et al., 2017; Joel and Ebenezer, 1993). Our survey findings revealed that 92% of the participants used the hook-and-line method when fishing. The emergence of the Thoothoor shark fishing fleet led to a significant increase in shark catches in Tamil Nadu between 1986 and 1997 (Kasim and Vivekanandan, 2011). The exact contribution of the Thoothoor fleet to this hike is unclear because the male fishers landed their catches at landing centres along the coast, including Kochi and other harbours in Kerala.

By 2000, the fleet had diversified into fishing for tuna for the local markets. The fleet comprised two distinct groups: boats with longlines for catching yellowfin tuna and boats with gill nets for catching skipjack tuna. More recently, the fleet has comprised three groups (Vivekanandan and Kasim, 2011):

1 boats with longlines used for shark fishing
2 boats with longlines and gill nets for catching tuna
3 boats with gill nets for catching seer fish and carangids.

The most interesting aspect of the fleet is its constant evolution, as it innovates and changes strategy according to the fishing opportunities. The fleet is estimated to have consisted of 588 boats in 2012 (Meenakumari,

2014). The boats are now larger and mechanized and have been fitted with upgraded equipment such as automatic identification systems (AIS), which enables them to fish long distances from the shore (see Endnote 3). Simultaneously, the costs have escalated, each boat costing between INR5–10 million (USD70,380–140,760).

Both the government of India and the government of Tamil Nadu expected the introduction of modern techniques of production in fishing, especially the motorization of fishing vessels and the boom in trawling, to yield significant economic benefits to the fishing communities (Johnson and Bavinck, 2010). While there have been gains, a new set of problems has emerged: a decline in fish catches due to intensification of fishing by motorized vessels (Bennett and Arumugam, 1993), a decline in the per capita income of small-scale artisanal fishers, and growing conflict between the territories occupied by traditional fishers and fishers who practise trawling, due to the finite nature of marine resources (Subramanian, 2009) (on the growing precarity in marine fisheries in Tamil Nadu, see Chapter 2, this volume).

## Migration and identity: bearing risks and learning skills

The growing competition for fish and the infrastructural difficulties in the study locality, such as the absence of harbours, has been a major driver for male Thoothoor fishers to fish off other parts of the west coast and in deep waters (Vivekanandan and Kasim, 2011). In our study, 94% of fishers reported such practices, and more than 80% of them were multiday fishers. Their needs for higher incomes to pay their debts, support their families, and fulfil their aspirations for themselves and their children are partly behind their seasonal movements to various destinations (David, 2013; Krishnan, 2010). In this respect, the male fishers' caste and kinship networks have served as a way to ensure employment and group mobility (for further discussion, see Chapter 9, this volume).

There is a further moral and personal dimension to the Mukkuvar fishers' mobility. Given the cultural connotations of fishing and the sea, they see the harmful effects of trawling on the environment and economy as an attack on divinity rather than as just an expression of the unequal distribution of resources (Kurien, n.d.; Subramanian, 2003). Hence, their movement to distant waters is a way of respecting nature and demonstrating their masculinity (Ram, 2008) by braving the harsh environmental conditions of those waters (Kennedy and Sundara Raj, 2014). In this respect, masculinity does not refer to an individual trait but rather to a way of ordering social practices, including divisions of labour, having access to wealth, resources and power, and the symbolic representations of meaning (Connell and Messerschmidt, 2005). In the process, the fishers ensure economic stability and simultaneously retain their distinct heritage.

### *Risks and forbearance*

Among the risks faced by the male Mukkuvar fishers, two need to be emphasized, both of which demonstrate their personal valour and capacity for endurance: environmental risks and social and/or personal risks. In this respect, notions of risk are interconnected and mediated by the history, location, and experience of the particular fishers (Power, 2008, p. 566). Storms and accidents are commonplace and have affected most households in Kanyakumari District. The most recent storm was Cyclone Ochki in December 2017, which left more than 150 people dead in the Thoothoor cluster, 11 of whom were from our study villages.[5] Boats and property are frequently destroyed by bad weather at sea or by accidents, and several participants mentioned the difficulties of securing compensation to rebuild their assets.

The physical risks faced by men at sea are mirrored in the women's capacity to bear uncertainties in the knowledge that fishing is their only source of income. As Tony's wife, Jessica,[6] said,

> Once they leave the shore, they are ours only if they come back. The women in the village pray for the men who have left for the sea. Each time there is news of a storm or accident, we panic, fearing for our men, yet we all bear this, as fishing is the main source of income. We cannot run a family if our men don't take that risk.[7]

During the FGDs and interviews, a number of men recounted their frightening experiences of storms and accidents at sea, when they could not cook or eat for days due to the high waves, when signals failed and boats collided, when there was loss of fishing vessels due to rocks and rough seas, or when they witnessed their friends die. Additionally, men face arrest when they drift into foreign waters. Tony's boat was captured by coast guards off the Maldives, and a few of his crew were arrested when he unknowingly entered the Maldives' territorial waters. Apart from the risk of survival, uncertainties are associated with economic costs, such as the purchase of a new boat, securing the release of a colleague from imprisonment after entering foreign waters without the proper permit, and making insurance payments, all of which affect the male fishers' livelihoods and well-being. However, as Jessica noted, the higher incomes meant the risks were worth taking (for an econometric discussion of the profitability of deep-sea fishing in Kanyakumari, see Chapter 12, this volume).

Most of the participants pointed to a range of health risks that they faced at sea. For example, Freddie recounted the rationing of water to 20 litres per week for bathing and washing. There are no toilets onboard, so the fishers use an open closet attached to the side of the boat. In the event of storms, it becomes difficult to use the facility, and then poor hygiene contributes to both discomfort and illnesses. If someone onboard falls sick, efforts are made by the crew to send the sick person ashore when the crew see a boat returning. However, Jasmine noted that many men had died at sea because, due to all the investments in the trip and the distance travelled, often over 1000 nautical miles, it was not possible for a boat to return without a catch. After the death of her neighbour, age 35, Jasmine had wondered how his wife would manage to raise their son, age 8. Nick shared his record of expenses with us, which clearly showed that the costs of fuel, groceries, ice, and labour for a two-week trip totalled a minimum of INR200,000 (USD2877.50). Hooks, longlines, and nets were additional costs.

During the summer, the men at sea are affected by the extreme heat. Food is not consumed regularly, especially when work is at a peak, such as when setting the nets and lines or hauling them back onboard (Sundar, 2018). Jose, the captain (*srangu*) of a boat, said he did not leave the controls, because the safety of the boat and those onboard depended on him, and often he went without food. Furthermore, while fishing, the crew have less sleep, sometimes two to three hours at a time. It can take hours to haul in the nets and lines, and crew cannot rest during that time, regardless of the weather. Many fishers chew betel leaves to keep themselves awake and warm, such as when temperatures drop at night. Betel leaves are also applied to wounds, particularly cuts, resulting from fish stings, sharp fish teeth, hooks, and ropes, because pulling in nets and anchors can tear the skin from the crew members' hands or cause joint pain.

The ability to cope with risk and uncertainty at sea means the fishers are tempted to engage in risky social behaviours, especially the consumption of liquor when on shore. This is opposed by the women, as heightened drinking leads to quarrels and violence in the family, apart from reducing the incomes available for running the household and managing debt. Devi said that her husband consumed liquor with the money he earned, leaving little for meeting the family's needs. She dreaded his arrival at home because he was also violent. She said, 'In my body, there is no place without injury. Probably no woman in this village has experienced such violence and abuse.' On such occasions, there was no peace in the house, and she felt that her husband's abusive behaviour had driven her eldest son to die by suicide. According to Clara, while the men did not consume alcohol during their fishing trip, women needed to fight the men's drinking habits at home because otherwise it could destroy families. In such cases, the women are supported by the local pastor and their neighbourhood *anbiam*, as the male manifestations of coping with risk and hard labour have the potential to destroy the gains in income that have contributed to the practice of multiday deep-sea fishing.

## Learning and developing skills

Deep-sea fishing for several days or weeks at a time involves hard work, which raises the question of what skills and knowledge are required. Rather than any formal training, the practice appears rooted in the Mukkuvar culture, ways of living and experience at sea, and possibly acquired through forms of 'apprenticeship' with more experienced fishers such as their fathers, caste elders, or friends (Rao and Hossain, 2012; Sundar, 2018). During a focus group discussion with six young men, the men revealed that they started fishing at sea when they were as young as 12. One of them said,

> Every day before going to school, early in the morning, we would play in the sea water, and then bathe in the river to remove the salt from our body. The gentle waves of the sea made us love fishing and the rough waves taught us to face the difficulties of life. At weekends, we would sometimes go on fishing trips with our fathers. Life at sea is tough, but our experiences from childhood have taught us how to tackle this.

When the male fishers were young, many who came from poor households dropped out of school after primary level or early secondary level in order to go fishing with their kin. Interestingly, among the older and the younger generations in the surveyed households in the study villages, more men dropped out after receiving primary or secondary education than did women (Table 7.1). While not formally taught the skills involved in fishing, the various dimensions of fishing are inculcated by playing in the sea and on the shore or by observing the elders performing tasks on the boats and, for example, talking about the sun and the wind. The male fishers' knowledge seems to be gained through observation and the shared performance of daily activities (Marchand, 2010). This highlights a tacit and embodied form of knowledge 'learned through non-formal processes and through relationships

*Table 7.1* Education levels of all adult members of households in the study villages, Kanyakumari District, Tamil Nadu, 2017

| *Education* | *Thoothoor village cluster* | | | | | | | |
|---|---|---|---|---|---|---|---|---|
| | *Men* | | | | *Women* | | | |
| | *19–40 years N = 93* | *%* | *41–60 years N = 59* | *%* | *19–40 years N = 92* | *%* | *41–60 years N = 53* | *%* |
| Illiterate | 0 | 0 | 7 | 11.9 | 0 | 0 | 5 | 9.4 |
| Primary | 19 | 20.4 | 32 | 54.2 | 2 | 2.2 | 10 | 18.9 |
| Upper primary | 14 | 15.1 | 13 | 22 | 8 | 8.7 | 16 | 30.2 |
| High school | 15 | 16.1 | 4 | 6.8 | 24 | 26.1 | 18 | 34 |
| Secondary school | 5 | 5.4 | 2 | 3.4 | 10 | 10.9 | 3 | 5.7 |
| Vocational | 9 | 9.7 | 0 | 0 | 6 | 6.5 | 0 | 0 |
| Graduate | 12 | 12.9 | 1 | 1.7 | 31 | 33.7 | 1 | 1.9 |
| Postgraduate | 3 | 3.2 | 0 | 0 | 7 | 7.6 | 0 | 0 |
| Professional | 16 | 17.2 | 0 | 0 | 4 | 4.3 | 0 | 0 |

Source: Household survey, 2017

of kin and place-based community' (Sundar, 2018, p. 16), which is a kind of experiential apprenticeship involving a host of everyday activities, often mundane work routines, rituals, and talk (Rao and Hossain, 2012).

However, not all of the male fishers in our study had learned at the same pace: some had been faster than others to learn a host of tactics and strategies, depending on their level of exposure. Before the year 2000, the male fishers identified areas with fish mainly by the flow and colour of the water and the position of the stars. However, since then, new technologies such as GPS (global positioning systems) and echo sounding have made it easier for them to locate places where fish are likely to be found, to monitor the course and speed of the boats, and to check the underwater terrain for rocks or other obstacles, thereby reducing risk and enhancing preparedness. Most of the elders were still able to estimate the sea level and conditions without modern equipment, and particularly during emergencies, when the equipment sometimes failed, their skills were useful for helping them reach the shore safely.

A further element of learning is the ability to innovate. Albert, a hard worker, had invented a new method, which involved using his main boat to tow his *vallam* (boat with an outboard motor) into deep waters. It had enabled him to take twice the amount of fishing equipment and use both the *vallam* and the main boat to set the lines in two different places 1–1.5 km apart and parallel to the boat (Box 7.1). With this method, Albert was able to reduce his time spent at sea and take fish back to shore while it was still fresh, which increased his income. Given the profitability of Albert's new method, other male fishers had started using it after 2012, by keeping a spare *vallam* to catch and store fish. They were also able to pay their crew more regularly and thus give them additional incentives to work hard.

### Box 7.1 A page from Albert's fishing diary

#### 28.09.2017

4.00 a.m.: Everyone woke up, had coffee, and started to prepare the fishing tools.

5.30 a.m.: We reached a place 14 nautical straight into the sea from Kochi.

6.00 a.m.: We started laying our line in the sea. The *vallam* went eastward, and we went westward, choosing entirely different places to get a higher chance of finding fish in either spot. Their line was laid 15 nautical miles away from us.

7.00 a.m.: As we were laying the line in the sea, we saw a few other *vallam* fishing for smaller-size kingfish [*Scomberomorus cavalla* (*S. cavalla*) or king mackerel]

7.30 a.m.: We finished laying the line and on the way got five small fishes using the plastic lure. The *vallam* did not get any in their plastic lure.
9.00 a.m.: We had breakfast and decided to use the plastic lure again because there was not much luck with the main line.
11.00 a.m.: We decided to pull out the line from the sea and use the drag lure for the rest of the day because the main line was hopeless that day.
1.30 p.m.: We pulled up our line completely and found a few small kingfish in it. The *vallam* also got a few small kingfish.
3.30 p.m.: When we were lure fishing, we did not eat together, but we took turns using the fishing lure, eating when we got time.
6.00 p.m.: We tried the drag lure the whole evening and got only a few fish. We could not even cover the expenses of the diesel used. So we called the *vallam* and tied it to the boat.
6.30 p.m.: We started towards Kochi harbour. I believed there will be squids available near the anchored ships close to the harbour and asked the crew to prepare the hooks for squids.
10.00 p.m.: The crew started to eat in turns and continued fishing.
11.00 p.m.: We got only half a basket of squid fishing till 11 p.m., and I asked to crew to stop for today. The crew went to sleep.

**29.09.2017:**

5.30 a.m.: We woke up, had coffee, and untied the *vallam* to let them fish.
6.00 a.m.: We let one *vallam* use the lure, and one was still tied to the boat. The boat did not run, because we did not want to waste more diesel.
9.30 a.m.: We had our breakfast and the running *vallam* called on wireless to inform us they were getting fish, so we started to move towards them.
11.00: On the way, we got five fish, and the *vallam* got four. Once we reached the *vallam*, we found they had caught 22 fish. They did not even have their breakfast, because they were busy fishing. So we prepared food and gave it to them.

### *Fishers' identity and the mobility of labour: growing tensions between wealth and precarity*

Given their reputation for shark hunting, Thoothoor fishers pride themselves on their skills and creativity. While there was not much distinction between the crew and the owners at sea, except perhaps for the special role and responsibility of the captain, most of the male fishers aspired to be boat owners and to build additional reputations on notions of honesty and success. Merlin was proud to be the first man to own a *vallam* in Thoothoor, which was entirely as a result of his hard work from a young age. Albert was recognized not only for his innovative method of towing an extra boat when fishing but also for his honesty and good character. He had a good reputation in the fishing industry. He was polite to everyone and had never cheated anyone, and therefore, others considered it safe to conduct business with him. Conducting business with Albert was also regarded as profitable because he was known for thinking, learning, and innovating. Similar to Albert, who had been fishing from an early age, Nick had been fishing from age seven. He owned two boats, from which he derived his identity. It had been his dream to own a boat, and he had worked hard to fulfil it. As a result, in his retirement, he was the secretary of the church committee and took part in prayers and other family and community events. His hard work had paid off, and he considered that he lived a good life as a consequence. Thus, his success in fishing had translated to a sense of overall well-being – material, relational, and subjective (for a definition of social well-being, see Chapter 1, this volume).

Despite their successes, given the high risks at sea and accompanied by declines in fish catches and growing competition, many of the older men wanted their sons to move into other professions. A further reason was the perceived market inequalities, especially in Kochi harbour, where the auctioneers pledged to their potential sons-in-law a fixed amount of INR2500 (USD35) per day from the catches as dowry (*streedhanam*). In a focus group discussion, several men noted that even boat owners and crew might not earn that amount, nor might the labourers engaged in loading and unloading the boat. However, the boat owners did not protest, because they believed that questioning the practice might create problems for them when landing their boats at Kochi harbour. Encouraged by the Church to further their education, many of the young men with educational or professional qualifications were seeking jobs elsewhere. By contrast, only 12.5% of the migrant fishers were planning to move to find professional jobs rather than fishing jobs. The results of our survey also indicated an intergenerational change given that only 50% of men in the age group 19–40 years engaged in fishing, compared with 93% of men over age 40.

Those who persist with fishing want to retain their identity and reinforce their masculinity as the most skilled among all male Indian fishers through their innovations, search for new fishing grounds, adoption of new techniques and technologies, boat ownership, and reputation. When they return home, they seek to portray a modern persona by using costly mobile phones, clothes, and hairstyles and by investing in a range of equipment and facilities. They use social media to post pictures and videos taken while working at sea.

We constructed a wealth index based on different elements of material status – housing, access to water and sanitation, educational levels, asset ownership, and employment – and unsurprisingly found that Thoothoor had less than 1% of households in the poorest category (Figure 7.1). Further, migrant workers, the majority of whom were multiday deep-sea fishers (see Table 9.1 in Chapter 9, this volume), confirmed during in-depth interviews that they were better off than non-migrants (as discussed in more detail in Chapter 12, this volume). Albert's wife noted,

> As he is away from home, on the deep sea, for long periods of time, the money comes with a delay, but in larger amounts. I have been able to save in chit funds and jewellery, have educated our children, built a new house, and bought land.

Not all young men can find professional jobs, and therefore, many shift from fishing to allied activities. Jose sold fishing equipment such as hooks, ropes, weights, floats, and food rations to deep-sea fishers. Additionally, he made sea anchors, which locally are called parachutes because they fill with water once dropped into the sea, and then water pressure holds them in place. Earlier, sea anchors were imported from Sri Lanka and were

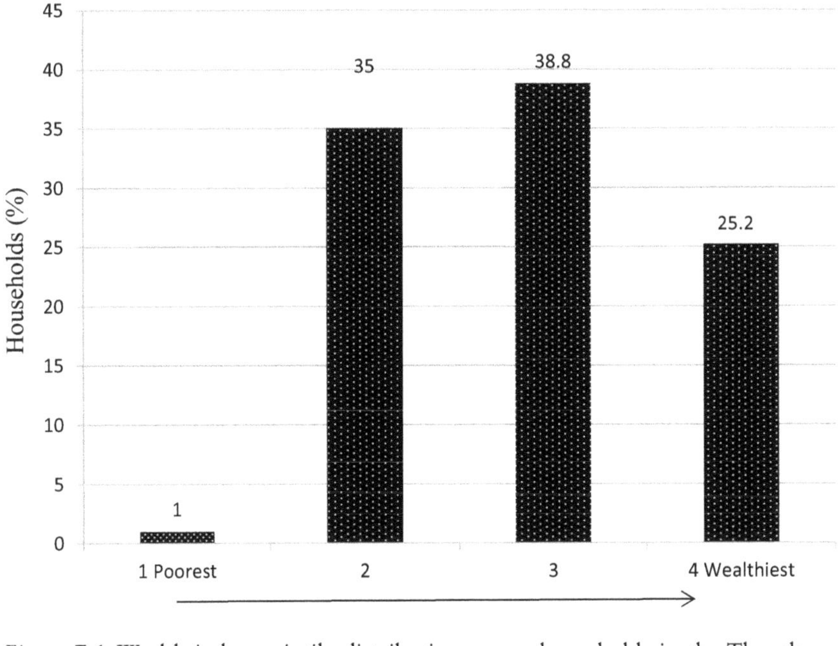


*Figure 7.1* Wealth index quintile distribution across households in the Thoothoor village cluster, Kanyakumari District, Tamil Nadu, 2017

Source: Household Survey (2017)

much more expensive. Binny adjusted and sold fishing equipment and ran a revolving financial management business. He said that most transactions were based on credit and depended on relationships of trust.

However, the decline in the numbers of young men taking up fishing has created a gap in the labour supply, which is now filled by migrant labour from distant locations in northern and northeast India (for a discussion of in-migrants into marine fisheries in Tamil Nadu, see Chapter 2, this volume). The migrants work alongside the local male fishers and are not discriminated against on the basis of their caste, religion, or social identity, due to the importance of teamwork. But there is no social mixing and the relationships remain purely professional. Many fishers emphasized that they allocated only non-professional activities such as storing the fish, pulling in the lines, or other minor tasks to the migrant workers, while they themselves did the skilled work such as calculating the point at which to set the lines on the basis of water movements or finding the best markets for their catch. Jose said, 'People who are not fishermen by birth will not be able to learn a few things about fishing. They can only see a few fish. They will not be able to learn the direction of the water current.'

Despite a discourse of equal shares, the migrant workers noted that they were paid less than other workers and that this was justified by the differences in skills, not just the practical skills that they had acquired through experience but also the less visible skills of observation and intuition that they were seen to lack. In this respect, the idea of 'specialization' as a common technique for maintaining wage gaps in globalized labour markets is clearly visible (Standing, 2011). While some migrant labourers stayed onboard the boat until the next trip, others stayed in separate rented housing and returned home once per year in June and July during the period when there was an annual ban on fishing. The migrant workers lack the identity of fishers, formally and informally, and hence lack any entitlements in terms of compensation for losses incurred at sea, including death. As one of them, Sunil, reported, four migrant workers died during cyclone Ochki, yet none of them received any compensation, or even a decent funeral. Furthermore, they did not receive compensation or rights to food in the locality during the ban on fishing, which in turn contributed to new forms of vulnerability and precarity (for further discussion on precarity, see Chapters 1 and 2, this volume). In this regard, the Mukkuvars' fisher identity carried more weight than the citizenship rights of migrant crew members with respect to differential access to benefits and services.

## Gendered identities and well-being

The foregoing discussion represents a masculine view of the deep-sea fishers of Thoothoor, and the masculinization of the industry has been found in earlier research (Power, 2008). Although women were in the past responsible for post-harvest operations when the fishing boats landed on the shore, ranging from the unloading of fish to marketing or drying, few women were engaged in the fishing enterprises when the survey was conducted. This was partly due to the decline in beach seine operations and the lack of landing facilities for the deep-sea fishing boats in the village, most of which continued to operate from harbours in Kerala. It was also due to the pride and identity of the deep-sea fishers as successful providers. Only widowed women or those with alcoholic husbands who were unable to provide for them engaged in fish vending. However, with men staying away from home for long periods of time, women's support had become even more central to male success, albeit in different ways. For instance, Albert's wife took care of all the finances and accounting tasks, including calculating shares and wages for the crew on his boat. She provided detailed statements of income and expenditure, and helped to recruit reliable labourers from the village when needed. Although she did not earn an independent income, she controlled all of Albert's income and used it to accumulate wealth and improve their joint well-being. Most of the women supported the fishing enterprises by extending their domestic labour, by performing tasks that were not readily visible in the public domain. Jasmine told us, 'I help my husband by fetching oil and kerosene from the utility shop. Fishing is a difficult task, and after work, men need to rest.'

Apart from supporting the fishing enterprises in indirect ways, women bore the entire responsibility for running the household and for supporting their children's education and marriage, ensuring financial stability and meeting the daily needs of their household, because the men were most often at sea. Women's responsibilities were reflected in their patterns of decision-making. They had a say in household matters, including household purchases, savings, and credit, while the men were increasingly dominant in decisions related to buying fishing equipment, migration, or doing other paid work (Figure 7.2). Not only is the way of life changing, but also the gendered nature of labour and its division is changing, as similarly found by Hapke in a study of Kerala (Hapke, 2012). In this regard, while fishing delineates men's work, their contributions to the household, and their identity, their advancement in the livelihood depends on having supportive wives who take responsibility for the management of the domestic domain (Power, 2008).

Mary's husband gave her all of his earnings and expected her to manage and spend the money well and to save some if possible. However, given the risky nature of fishing as a livelihood and the associated uncertainties, a regular income is not guaranteed, thus making women's tasks harder (see Endnote 4). While Mary prided herself on being one of the best mothers in her village, she was devastated when she learned that her eldest daughter had a romantic relationship with a boy from their village: 'Children should understand that parents who brought them up also have the knowledge to get them a good partner. I considered myself a failure as a mother that day.' She regarded her daughter's decision as a result of her own inability to

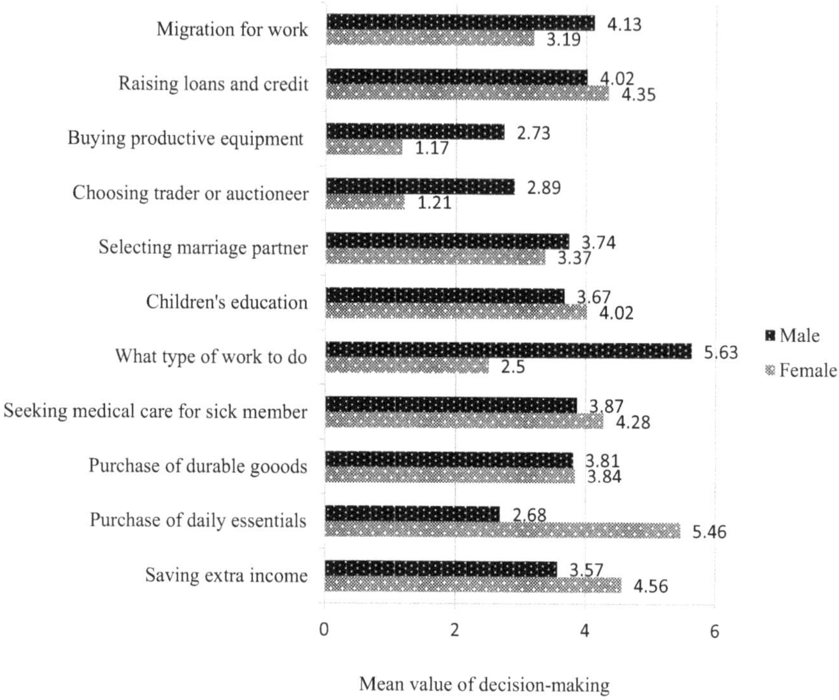


*Figure 7.2* Gendered decision-making in the Thoothoor village cluster, Kanyakumari District, Tamil Nadu, 2017 (mean values)

Source: Household Survey (2017)

manage the domestic domain effectively, which thus reflected negatively on her skills and identity as a mother.

Other identities can redress the imbalance in women's lack of control over production and to some extent the domestic domain, especially as educated children are tending to move in order to work as professionals in different locations, and in the process, they are exercising choice in selecting their marriage partner, place of residence, or engagement in work. Although professional employment is not encouraged, with the possible exception of teaching, and women's roles as carers are valorized, Jessie was proud of her identity as a Christian. Like Mary and most of the other women in the study villages, she was an active member of the Church and *anbiam*, and she engaged in a range of social and political events. She prioritized her Church activities over her belonging to the Mukkuvar caste, considered low in the caste hierarchy, apart from carrying connotations of risky behaviour as a fishing caste. For the women, well-being was derived relationally from the success of their husbands and subjectively from their own sense of

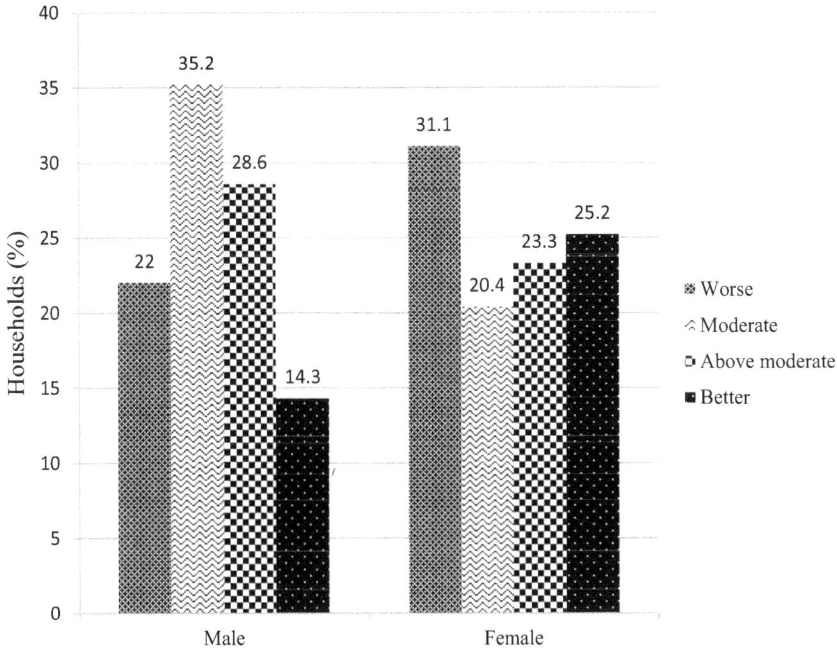


*Figure 7.3* Men's and women's perception of well-being in 2017 compared with in 2007, Thoothoor village cluster, Kanyakumari District, Tamil Nadu

Source: Household Survey (2017)

contribution to the community (through church activities), if not from the household.

We asked people in households in the study villages a series of questions relating to their perceptions of well-being compared with a decade beforehand. Their responses were scaled and grouped into quartiles, ranging from a perception of being worse off to a perception of being much better off in their well-being (Figure 7.3). Interestingly, more women than men reported a positive change – materially through their ability to accumulate wealth, socially through their involvement in church-based and *anbiam*-led social and political activities, and subjectively in a sense of status and respect in the community. However, more women than men also reported being worse off and pointed mainly to the wide prevalence of alcoholism and other processes of social exclusion at work.

## Conclusions

Apart from resource depletion and changing fishing practices (mechanized fleet, increased near-shore competition), it is clear that multiday, deep-sea fishing by the Mukkuvars of Kanyakumari District is driven by a discourse of skills and professional pride as fishers. For the satisfaction of being 'shark hunters', they are willing to travel hundreds of kilometres to the deep-sea fishing areas, at considerable risk to their life and health and with uncertainties about catch and earnings. The fishers' demonstration of their creativity and ability to find and catch sharks, and more recently tuna, contributes to the fulfilment of their identities both as men and as members of a particular caste. In this respect, their skills and identity need to be understood in the context of their caste identity and the social relations and discourse in which they are embedded.

Although women were formerly central to the fishing enterprises, especially post-harvest operations, they are now seen as homemakers, responsible for the care of the family and reproduction of the household, especially when the men are away. Educated women are not expected to engage in fish vending, which is regarded as low status work, or in other paid work or employment; rather, their contributions are restricted entirely to the domestic domain. The domestic domain has been redefined to include a range of supportive activities that are critical to the fishing enterprises, but although this expands the sphere of women's agency, it is not valued in economic terms (Rao, 2018).

Stories of fishers' bravery and self-actualization, as well as their potential earnings, show that fishers are willing to face a range of risks, which they justify in terms of the welfare and well-being of their families. However, urgent attention is needed to find ways to minimize the risks through forms of co-management that could enhance their safety. Ultimately, the discourse of risk is used to mediate and maintain structures of inequality, including those of gender. The fishing sector can respond to outstanding needs and ensure the well-being of the fishers taking high risks on the coast of Tamil Nadu only if fisheries management policies cover issues of identity and social reproduction in addition to those of production.

## Notes

1 The term *panchayat* refers to the lowest tier of elected government, which operates at the level of a cluster of villages.
2 Ancient Tamil literature composed by poets (some of them anonymous) from southern India and dating from the period c. 300 BC to AD 300.
3 Unpublished report titled 'Coastal population dynamics and ecosystem changes: How markets, technology and Institutions affect this process along the West Coast of India' by N. Nayak, D. Nandakumar and A.J. Vijayan, 2006, Protsahan.
4 Unpublished report titled 'Shark Fishers' Summit: Proceedings and recommendations. Association of Deep Sea Going Artisanal Fishermen' by J.V. Jain and S. Carmel, S., 2002, for international circulation in the Association of Deep Sea Going Fishermen.
5 Unpublished report titled 'The cyclonic apartheid' by the People's Inquest Team in Kanyakumari District, 28–29 December 2017.
6 All participants' names are pseudonyms.
7 All quotes from the study participants were translated from Tamil by the authors of this chapter.

## References

Abobi, S.M., and Alhassan, E.H. (2015). A review of fisheries-related human migration in the Gulf of Guinea. *Journal of Coastal Zone Management, 18*, 395. doi:10.4172/2473-3350.1000395

Acheson, J.M. (1981). Anthropology of fishing. *Annual Review of Anthropology, 109*, 275–316.

அகநானூறு [*Agananooru*]. (n.d.). [Anthology of Sangam literature, 7th book]. www.tamilsurangam.in/literatures/ettuthogai/agananooru/index.html accessed 5/2/2020.

Bavinck, M. (2014). Investigating poverty through the lens of riches – Immigration and segregation in Indian capture fisheries. *Development Policy Review, 32*(1), 33–52.

Bennett, S.P., and Arumugam, G. (1993). *Impact of motorisation on the traditional fishery of Tuticorin*. Marine Fisheries Information Services, Technical and Extension Series No. 123. Retrieved from http://eprints.cmfri.org.in/3497/

Binkley, M. (1995). *Risks, dangers and rewards in the Nova Scotia offshore fishery*. Montreal: McGill-Queen's University Press.

Connell, R.W., and Messerschmidt, J.W. (2005). Hegemonic masculinity: Rethinking the concept. *Gender and Society, 19*(6), 829–859.

David, G. (2013). Constraints of outmigrant fishermen diaspora in Kanyakumari. *Indian Streams Research Journal, 3*(6), 1–2. doi:10.9780/22307850

Fernando, H., Anandhan, P.S., and Daniel, N. (2017). Shark fishing – A unique traditional fishing practice in Thoothoor, Tamil Nadu, India. *Ecology, Environment and Conservation, 23*(September Supplementary Issue), 10–15.

Hapke, H. (2012). Capturing the complexities of globalization in fisheries: Gendered divisions of labour and difference. *Asian Fisheries Science Special Issue, 25S*, 75–92.

Hapke, H., and Ayyankeril, D. (2018). Gulf migration and changing patterns of gender identities in a South Indian Muslim Community. In M. Amrith and N. Sahraoui (Eds.), *Gender, work and migration: Agency in gendered labour settings* (pp. 175–192). New Delhi: Routledge.

Hornell, J. (1908). *Madras fishery investigations*. Madras Fisheries Bulletin 4.50. India Office Records and Private Papers archive, British Library, London.

Joel, J.J., and Ebenezer, L.P. (1993). *Long lining specifically for sharks, practised at Thoothoor*. Marine Fisheries Information Service No.121. Retrieved from http://eprints.cmfri.org.in/3411/1/Article_04.pdf accessed 5/2/2020.

Johnson, D., and Bavinck, M. (2010). *Social justice and fisheries in India, social justice and fisheries governance: The view from India*. Rome: FAO.

Kasim, H.M., and Vivekanandan, V. (2011). *Marine fisheries production in Tamil Nadu and Puducherry*. Work Package 5, Report – WP5AR2, Fisheries Management for Sustainable Livelihoods (FIMSUL) Project. Retrieved from http://eprints.cmfri.org.in/9253/1/35.pdf

Kennedy, J.D., and Sundara, R. (2014). Determinants of fishermen migration: Study on a traditional marine fishing village in Kanyakumari District. *Acme Intellects International Journal of Research in Management, Social Sciences & Technology, 8*(8), 2320–2793.

Krishnan, V. (2010). *An economic study on migration of fishermen from Kanyakumari district*. Doctoral thesis, Manonmaniam Sundaranar University, Tirunelveli.

Kurien, J.T. (n.d.). *People and the sea: A 'tropical-majority' world perspective*. Retrieved from www.marecentre.nl/mast/documents/JohnKurien.pdf accessed 5/2/2020.

Marchand, T.H. (2010). Making knowledge: Explanations of the indissoluble relation between minds, bodies and environment. *Journal of Royal Anthropological Institute*, *165*(1), S1–S21.

McGoodwin, J.R. (2001). *Understanding the cultures of fishing communities: A key to fisheries management and food security*. FAO Fisheries Technical Paper 401. Rome: FAO. Retrieved from www.fao.org/3/y1290e00.htm accessed 5/2/2020.

Meenakumari, B. (2014). *Report of the expert committee constituted for comprehensive review of the deep sea fishing policy and guidelines*. Retrieved from www.researchgate.net/profile/Meenakumari_Bharathiamma2/publication/322821544_Report_of_the_Expert_Committee_constituted_for_Comprehensive_Review_of_the_Deep_Sea_fishing_Policy_and_Guidelines/links/5ab4def3aca2722b97c9783d/Report-of-the-Expert-Committee-constituted-for-Comprehensive-Review-of-the-Deep-Sea-fishing-Policy-and-Guidelines.pdf

Moore, H.L. (1994). *A passion for difference*. Cambridge: Polity Press.

Narendran, V. (2016). *A study on migration of fishers from Kanyakumari to the Cooperation Council for the Arab States of the Gulf (CGASG)*. ICSF Occasional Paper. Retrieved from www.icsf.net/images/occasionalpapers/pdf/english/issue_156/156_Kanyakumari%20study_occasional%20paper_27_FINAL.pdf accessed 5/2/2020.

Pollnac, R.B., and Poggie, J.J. (1988). The structure of job satisfaction among New England fishermen and its application to fisheries management policy. *American Anthropologist*, *90*, 888–901.

Pollnac, R.B., Poggie, J.J., and Cabral, S.L. (1998). Thresholds of danger: Perceived risk in a New England fishery. *Human Organization*, *57*(1), 53–59.

Power, N.G. (2008). Occupational risks, safety and masculinity: Newfoundland fish harvesters' experiences and understandings of fishery risks. *Health, Risk and Society*, *10*(6), 565–583.

Ram, K. (1991). *Mukkuvar women: Gender, hegemony and capitalist transformation in a South Indian fishing community*. London: Zed Books.

Ram, K. (2008). The Mukkuvars of Kanyakumari on the margins of caste society. In I. Banerjee-Dube (Ed.), *Caste in history* (pp. 136–149). New Delhi: Oxford University Press India.

Rao, N. (2018). Fertility, reproduction and conjugal loyalty: Renegotiating gender relations among dalits in rural Tamil Nadu. *South Asia Multidisciplinary Academic Journal*, *19*. doi:10.4000/samaj.4575

Rao, N., and Hossain, M.I. (2012). 'I want to be respected': Migration, mobility and the construction of alternate educational discourses in rural Bangladesh. *Anthropology and Education Quarterly*, *43*(4), 415–428.

Rigg, J., Oven, K.J., Basyal, G.P., and Lamichhane, R. (2016). Between a rock and a hard place: Vulnerability and precarity in rural Nepal. *Geoforum*, *76*, 63–874.

Sathiadhas, R., and Prathap, S.K. (2009). Employment scenario and labour migration in marine fisheries. *Asian Fisheries Science*, *22*, 713–727.

Standing, G. (2011). *The precariat: The new dangerous class*. London: Bloomsbury.

Subramanian, A. (2003). Community, class and conservation: Development politics on the Kanyakumari coast. *Conservation and Society*, *1*(2), 177–208.

Subramanian, A. (2009). *Shorelines: Space and rights in South India*. Stanford, CA: Stanford University Press.

Sundar, A. (2018). Skills for work and the work of skills: Community, labour and technological change in India's artisanal fisheries. *Journal of South Asian Development*, *13*(3), 1–21.

Swathilekshmi, P.S., and Johnson, B. (2015). *Migrant labourers in the primary sector of marine fisheries: A case study in Karnataka*. Central Marine Fisheries Research Institute. Retrieved from http://eprints.cmfri.org.in/9870/1/Swathilakshmi_6.pdf accessed 5/2/2020.

Torrente, F., Bambridge, T., Planes, S., Guiart, J., and Clua, E.G. (2018). Sea swallowers and land devourers: Can shark lore facilitate conservation? *Human Ecology*, *46*(5), 717–726.

Vivekanandan, V., and Kasim, H.M. (2011). *Fisheries management option for Tamil Nadu and Puducherry*. Work Package 5, Report R-20, Fisheries Management for Sustainable Livelihoods (FIMSUL) Project. Retrieved from http://eprints.cmfri.org.in/9251/1/33.pdf

# 8 Immobility of ethnic Vietnamese fishers and their adaptation strategy in Chhnok Tru village, Chhnok Tru Commune, Cambodia

*Raksa Sok and Kyoko Kusakabe*

## Introduction

Tonle Sap lake in Cambodia is a source of livelihood for millions of people in Cambodia, but fish catches there have been declining in recent years (FAO, 2013; Phanith, 2013), which in turn has impacted people's livelihoods. Phanith (2013) cites Serrat (2005) in attributing the decreases in catches to reduced water levels, unclear property rights for fishing, and hydrological changes. Although fishing lots were abolished in 2012, fish catches continue to decrease, and there is fierce competition over fish resources (FAO, 2013). In this chapter, we analyse the coping strategies of ethnic Vietnamese fishers who live in floating houses on Tonle Sap and who fish in the lake for their livelihood. Unlike the fishers described in Chapter 3, all of whom were Khmer, ethnic Vietnamese face different challenges, including restrictions on their mobility and access to alternative livelihoods.

There are several ethnic groups among the fishers who fish in Tonle Sap, including Vietnamese, Cham, and Khmer. According to Takizawa et al. (2009), cohesion between the Vietnamese and Khmer is not strong, not only because of their cultural differences but also because of the historical conflict between Cambodia and Vietnam. Many of the ethnic Vietnamese do not have Cambodian citizenship, and therefore, according to the 2005 subdecree on community fisheries management (Kingdom of Cambodia, 2005), they are not eligible to form or be members of community fisheries (CFs).[1] Hence, there are no CFs in Chhnok Tru floating village, where ethnic Vietnamese fishers constitute the majority of all fishers who live in the village. Further, ethnic Vietnamese people do not have rights to landownership (Minority Rights Organization, 2015, 2017), so they are forced to live in floating houses. It is difficult for them to migrate to other countries since they do not have legal documents to apply for safe migration. Some ethnic Vietnamese people do not speak the Khmer language and therefore find that communication outside the community is difficult, which also limits their income-generating opportunities. Compared with the Khmer people, ethnic Vietnamese face greater disadvantages – for example, their children do not attend school, not only because of their lack of Cambodian citizenship but also because they are required to help their poor parents with fishing (Jeanne et al., 2013).

In the following, we start with a description of the history of migration and the citizenship situation of ethnic Vietnamese in Cambodia, followed by an overview of the analytical framework used to understand immobility in the case of the ethnic Vietnamese. Thereafter, we describe the methods used in the study and then present an analysis of different factors that affect the ethnic Vietnamese people's adaptation strategies, given their limited mobility. The chapter is concluded with a summary of how flexibility in gender norms and equal gender relations are important resources for ethnic Vietnamese people to adapt successfully to decreases in fish catches.

## Ethnic Vietnamese on Tonle Sap

The exact number of Vietnamese living in Cambodia, including both recent and long-term migrants, is unknown, because many of them are undocumented (Kirchner, 2015). Even the Vietnamese who have lived in Cambodia for several generations remain undocumented (MIRO, 2015, 2017). According to Parsons and Lawreniuk (2018), there are approximately 500,000 ethnic Vietnamese people among the 15 million population of Cambodia, which means Cambodia has the highest proportion of stateless people in the world. This chapter focuses on ethnic Vietnamese people who lived in floating houses on Tonle Sap in late 2017 and early 2018. According to one estimate, approximately 300,000–500,000 people live in floating houses around the edges of the lake and Tonle Sap river (Kirchner, 2015). The houses are constructed on rafts made of bamboo poles with barrels attached below to keep them afloat. It is dangerous to live on floating houses, and children are always at risk of falling into the water. Winds and waves are strong on the lake, and therefore, the houses are easily and quickly damaged or broken. Those who cannot afford repairs continue to live in their damaged houses, which exposes them to greater risks (Kirchner, 2015). Many of the ethnic Vietnamese living on the lake do so because they are unable to own land and houses. Others live on the lake because they have a tradition of working as fishers and have always lived close to water bodies (Jeanne et al. 2013). Due to a lack of citizenship, ethnic Vietnamese people's mobility is restricted, and hence, there is little out-migration by the ethnic Vietnamese. This has led to an increase in the population on the lake and greater competition for fish resources.

According to Ehrentraut's history of the Vietnamese in Cambodia, the people have been migrating to the country since as early as 10th century (Ehrentraut, 2011). In southern Cambodia, the Vietnamese have occupied part of the lower Mekong Delta. The expansion of Vietnamese occupation in Cambodia ceased when the French colonized Indochina, but the French later encouraged Vietnamese migration to Cambodia as part of their occupation of the country. After Cambodia gained independence from the French in 1953, Prince Sihanouk declared Cambodia as the country of the Khmer people, and subsequently, Lon Nol's regime adopted an anti-Vietnamese policy that led to large-scale forced repatriation as well as the killing of ethnic Vietnamese. The Vietnamese were targeted for persecution during the Khmer Rouge's period of rule (1975–1979), and although approximately 30,000 ethnic Vietnamese escaped to Vietnam, 100,000 died in Cambodia (Kiernan, 1990). When the Khmer Rouge regime collapsed with the invasion of the Vietnamese army, the People's Republic of Kampuchea (PRK) was established. The PRK was pro-Vietnamese, and ethnic Vietnamese started to return to Cambodia after 1979. However, anti-Vietnamese sentiments were refuelled by the large number of Vietnamese in the country and the historical memory of both the invasion by the Vietnamese and the Vietnamese government's imposed administration of Cambodia (Amer, 2013; Chandler, 2008). Although the Comprehensive Cambodian Peace Agreements were signed in 1991, both the PRK and the Khmer Rouge reported a number of violent incidents directed towards ethnic Vietnamese in the 1990s (Takizawa et al., 2009). Such anti-Vietnamese sentiments were exploited by the Khmer Rouge until the regime was eliminated in 1998 and are still exploited by opposition parties in the current government (Amer, 2013).

## Immobility as disempowerment for ethnic Vietnamese on Tonle Sap

Ehrentraut explains how nation-building in Cambodia was based on 'thick national Khmer identity' (Ehrentraut, 2011, p. 784) and citizenship was defined by ethnicity. Following W. Kymlicka's classification (Kymlicka 1995, cited in Ehrentraut, 2011), Ehrentraut points out that as 'long-term residents who are denied citizenship', such as irregular migrants and refugees, the ethnic Vietnamese in Cambodia have the characteristics of 'metics' (Ehrentraut, 2011, p. 781). Ehrentraut (2011) calls those who are excluded from the mainstream even when they want to integrate as 'metics'. 'Metics' is a different category from national minorities, who claim the right to maintain their own identities, and from voluntary immigrants who aspire to full citizenship in a country, which they are often granted. Although the ethnic Vietnamese have lived in Cambodia for many generations, they are effectively stateless due to their lack of Vietnamese citizenship (Ang et al., 2014; Sperfeldt, 2017).

The liminal position of ethnic Vietnamese deprives them of voting rights, access to residential land and farming land, access to loans, access to education, access to formal employment, the freedom to travel in and out of country, the ability to obtain a driving licence, and access to law enforcement (Ehrentraut, 2011; Kirchner, 2015; Parsons and Lawreniuk, 2018). They also suffer from police harassment due to their precarious status and lack of legal documents (Minority Rights Organization, 2015). Although the 'thick national Khmer identity' excludes ethnic Vietnamese from citizenship, they can become naturalized if they wish (Nguyen and Sperfeldt, 2012). However, naturalization is difficult: some ethnic Vietnamese paid KHR250,000 (USD62.4) to immigration officers to obtain a Cambodian ID card (Kong Meta and Nachemson, 2017). Some Vietnamese previously had ID cards, but their cards were taken away from them when they returned after the Khmer Rouge regime. Parsons and Lawreniuk (2018) use the term *liminal statelessness* to describe the grey and fuzzy situation. They point out that such liminal statelessness keeps ethnic Vietnamese in a limbo, in that they are not able to move freely or to remain in the place. They are constantly vulnerable to eviction, and they need to live in ethnic enclaves to protect themselves from police harassment (Jeanne et al., 2013; Parsons and Lawreniuk, 2018). In their community, they can move freely, but they are not able to look for employment elsewhere (Kirchner, 2015; Minority Rights Organization, 2015). Those without a national ID card are unable to work in factories, and they cannot legally migrate to other countries.

Thus, ethnic Vietnamese in Cambodia have been left immobile, unable to move to look for other livelihood options, and consequently, their adaptation strategies can be limited. Immobility can be disempowering in other ways. For example, the poor are unable to move away from environmental threats and are therefore trapped and remain vulnerable to environmental changes, which led to decreases in fish resources (Afifi et al., 2016; Bettini and Gioli 2016; Black et al., 2011). Kaufmann (2002) calls for recognizing the importance of the concept of motility – the ability to be mobile – to better understand the situation of the most marginalized people in a society, as it is often the case that immobility is due to the lack of ability and resources. Furthermore, when there are few resources under restricted mobility, both gender relations and flexibility in gender norms can be considered resources in an adaptation strategy. Women's empowerment and equal gender relations lead to better economic outcomes (Haque and Kusakabe, 2005; Kabeer 2012; World Bank, 2011). In this chapter, we explore how the ethnic Vietnamese who are able to adapt to the decreases in fishing resources have demonstrated higher flexibility in gender roles and norms. Women in ethnic Vietnamese communities can be more disadvantaged than women in Khmer communities, because of their limited mobility resulting from gender-based divisions of labour and because of their lower levels of education and reduced access to resources such as finance, land, and information (Ha, 2008; LaBorde, 1996; Long et al., 2000; Shelton and John, 1996). However, in the absence of such resources, ethnic Vietnamese households can loosen the gender norms that restrict the gendered division of labour and thus adapt to crises in their livelihoods.

## Methods

The study was conducted from September 2017 to January 2018 in Chhnok Tru village in Kampong Chhnang Province. In-depth interviews were held to explore the coping strategies of ethnic Vietnamese people in the face of decreases in fishing resources and explore how their coping strategies have changed gender relations and vice versa. Part of Chhnok Tru village exists on water, where villagers live in floating houses. According to the police officer in Chhnok Tru (interviewed on 30 November 2017), the total number of households in 2016 was 1344, of which 516 were Khmer households, 48 were ethnic Cham households, and 780 were ethnic Vietnamese households. Our study focused on only the ethnic Vietnamese households in the village. Due to language barriers, only those who could speak the Khmer language and were willing to be interviewed were selected as study participants. The interviews were conducted in Khmer by the first author. Some ethnic Vietnamese were not willing to be interviewed, and therefore, the first author requested the head of the Vietnamese association in Chhnok Tru Commune, Sambor District, Kampong Chhnang Province (*Samakhom chong chiet Vietnam kanong khum Chhnok Tru srok sambor kaet kampong chhnang*) to introduce her to other potential participants. In total, 27 participants agreed to participate in the study: 14 women and 13 men. All participants were born in Cambodia, but none of them had Cambodian citizenship. Some had fled temporarily to Vietnam during the Khmer Rouge regime, but otherwise, they had lived in Cambodia all their lives. They said that their ID cards were taken from them by the local authority and that they were applying for residential immigrant cards. The highest level of education reported by participants was primary school level (grade 1) in a Vietnamese private school. None of them attended Cambodian public schools. Apart from the fishers, representatives of local authorities and an NGO (the Minority Rights Organization) in Cambodia were interviewed. In addition, two focus group discussions were held, with five participants in a women's group and six in a men's group. All of the men were fishers, while the women included fishers, homemakers, and vendors.

## Coping strategies of ethnic Vietnamese fishing households

As discussed in Chapter 3, fishers have experienced significant decreases in fish catches from Tonle Sap. The participants said that before the year 2000, they were able to catch 40–60 kg per day during the *prahok* (fermented fish) season (i.e. the season when they caught the most fish), but they caught only 20–30 kg in 2017. When the catches were too small and they were not able to make a living, they borrowed money from their parents and relatives and/or their neighbours. A few participants borrowed from fish collectors[2] and repaid them in fish. The Cambodian villagers can borrow money from micro-finance institutions in the village, but the ethnic Vietnamese are unable to do so, because they lack the required documents. Women seek other sources of income, such as selling things and cutting off fish heads,[3] to cover the shortfall in men's fishing income.

Faced with the depletion of fishing resources and without an option to move out of the village, due to limited mobility and citizenship rights, the ethnic Vietnamese fishing households had three options (Table 8.1):

1 They could continue to fish but have smaller catches.
2 They could continue to fish but also engage in other income-generating activities to supplement their income, such as by working in aquaculture (fish farming) or as daily wage labourers, as fish processers in households that made *prahok*.
3 They could stop fishing and instead earn a living in aquaculture, as daily wage workers or work in the retail trade. Those in this third group had the highest incomes compared with those in the other two groups.

Although all Vietnamese households in the floating village had limitations in terms of their citizenship, mobility, and options, there were differences in how they tried to cope. The differences in the households' coping strategies

*Table 8.1* Profile of the three groups of households in Chhnok Tru village, Chhnok Tru Commune, Cambodia, by response to decreases in fish catches, 2018

| | *Household group* | | |
|---|---|---|---|
| | *Group 1: Fishing only* | *Group 2: Fishing plus other activities* | *Group 3: Non-fishing activities* |
| Number of participants | 5 (3 men, 2 women) | 15 (8 men, 7 women) | 7 (2 men, 5 women) |
| Livelihood assets | Boat<br>Net | Boat<br>Net<br>Cages for fish farming | Boat<br>Cages for fish farming |
| Livelihood activities | Fishing | Fishing<br>Fish head cutting<br>Labouring<br>Small-scale retailing<br>Fish farming (cage) | Fish head cutting<br>Labouring<br>Small-scale retailing<br>Fish farming (cage) |
| Age of study participant (years) | 21–65 | 23–58 | 20–58 |
| Household size (number of people) | 2–4 | 3–7 | 2–7 |
| Number of employed household members | 1–3 | 2–6 | 2–6 |
| Education | None | 1 | 1 |
| Average income per day | KHR15,000–20,000 (USD3.75–5) | KHR30,000–40,000 (USD7.5–10) | KHR35,000–58,000 (USD8.75–14.5) |

Source: In-depth interviews, 2018

depended on factors such as household size and structure and their social network, gender norms, and relations. In the following sections, we explore each of these factors in turn.

### *Household size and structures as determinants of adaptation strategies*

One of the reasons why some households only did fishing, while others branched out to various other livelihood options, relates to the number of members in the household. If the numbers were small and there was only one person of working age, the household kept to fishing as a source of income. Furthermore, some elderly households without dependents decided to live on smaller income and therefore preferred not to change their fishing livelihoods. Mr Vuthy[4] said that he continued to fish for a living because he did not need much income:

> Yes, the depletion of fishing resources is really hard for some families, even my family. Since my family is smaller than before [his children had married and lived in separate households], I have to earn for only two people, so it is not so difficult for us.
>
> (Mr Vuthy, married, 65 years, in-depth interview, 2018; Group 1)

Mr Vichit said that the younger household members preferred non-fishing work, but elderly households continued to fish for a living: 'Many people in the village work as daily wage labourers, but they are mostly young people' (Mr Vichit, married, 54 years, in-depth interview, 2018; Group 1). By contrast, Mr Vath had a larger household and was able to branch out into other activities, although his family still experienced financial difficulties:

> Because income from fishing is not enough to cover family expenditure, my daughter and my wife decided to go [to work as daily wage workers] to cut fish heads in the village during *prahok* season to earn additional income. My son and I go fishing. Sometimes, he finds other wage work in the village to increase household income. I can earn around 20,000 riel [USD5] per day from fishing, and my daughter and my wife can earn around 8000 to 10,000 riel [USD2–2.5] per day during the *prahok* season. However, none of these incomes are permanent and sometimes we cannot earn any income.
>
> (Mr Vath, married, 56 years, in-depth interview, 2018; Group 2)

The households in Group 2 had more members and therefore needed more income for survival. However, the household's income from fishing was no longer enough to meet their requirements. Hence, following the depletion of fish resources, they had started to diversify their income within the village.

Further, with more working adults, they were able to find alternative livelihoods while the head of household continued to fish:

> My father and younger brother go fishing. However, since the catches are decreasing, I decided to sell vegetables on the boat around the village. The money from retail vending is not much, but it is better than depending only on fishing activity, as we need to feed two young children and my grandmother.
>
> (Ms Sophea, single, 23 years, in-depth interview, 2018; Group 2)

In most cases, the women sought non-fishing livelihoods. Some women started micro-businesses and worked as daily wage workers inside the village to supplement their household's income. However, such income was seen as small and insignificant, and fishing was still considered the main occupation of the household.

In households in which more members could engage in different income-generating work, diversification was employed as a strategy while still maintaining their fishing activities, which were considered their main livelihood. Thus, all other forms of income were considered supplementary:

> I have two sons. They are fishers. I mostly go fishing with them. However, the decrease in fish catches affected our household's survival, so I started keeping fish in cages since 2014 to earn more income. My wife is responsible for all housework, and sometimes she goes cutting fish heads in the village [for daily wage work]. The total income is not as high as before. Income from fishing is around 20,000 riel [USD5] and [with other income] the total will be around 30,000–40,000 riel [USD7.5–10] per day. We can earn around one million riel [USD250] from fish in the cage each year, but the income is earned only once a year. Life on the water is hard. We cannot depend on only fishing, since the number of fish has decreased.
>
> (Mr Sareth, married, 45 years, in-depth interview, 2018; Group 2)

### *Social network as a determinant of an adaptation strategy*

Due to the limitations in terms of the ethnic Vietnamese villagers' mobility and language, they normally have limited access to people and information outside their own community. Hence, they find that employment opportunities are extremely limited. Mr Naeth continued to fish with his wife and said that he did not have any experience in other occupations or any savings with which to start up new income-generating activities. Therefore, they continued to fish as a couple: the husband fished, and the wife paddled their boat:

> With fishing, we can survive from day to day, but the income is not enough to save for future investments, as the expenditure in my family

> is high. Anyway, there are few working opportunities for people living in floating houses.
>
> (Mr Naeth, married, 45 years, focus group discussion, 2018; Group 1)

Previous experience of running a business or learning skills from neighbours, friends, or other family members was a key factor in such decision-making:

> Due to the depletion in fishing resources, my income from fishing is not enough to support our household. We work harder than before, but we still cannot earn as much as we did earlier. I decided to stop fishing and started keeping fish in cages. I learned how to keep fish from a friend outside the village. My son and daughter-in-law sell vegetables in the village. Our current income is enough to make ends meet.
>
> (Mr Phally, married, 45 years old, in-depth interview, 2018; Group 3)

Another participant, Ms Sarom (aged 46 years; Group 3) started to sell noodles after she had learned from her mother how to cook noodle soup.

Although none of the participants were able to access formal loans, some were able to mobilize their social network in order to borrow money from collectors, family, friends, and/or neighbours, but the amounts were too small for investment purposes. Those who were able to invest in aquaculture or other income-generating activities had used their own savings from fishing.

### *Transformation of gender relations as an adaptation strategy*

We found that in households in which gender roles and norms were more flexible, the members tended to be able to adapt better to the decreases in fish resources. Mr Vichit expected his wife to stay at home and look after the household while he fished: 'I do not want my wife to work outside the household, because she has to do the household chores and take care of the children. These are her responsibilities because she is a woman' (Mr Vichit, married, 54 years, in-depth interview, 2018; Group 1).

The women in Group 1 managed all the household chores and also played a supporting role in the men's fishing activities. In cases where husbands were the sole income earner for the family, they tended to dominate decision-making in the household. The male participants in Group 1 were unwilling to change to other occupations, while the wives tried in vain to push their husbands to change:

> My husband is the head of the family. He has the power to make all the decisions in the household. Even after the decrease in fish resources, my husband decided to continue fishing. I asked him to find other work

> in the village, but he didn't agree. I know that the income from fishing is not enough, but he decided to continue fishing rather than do other work. He does not listen to me because he thinks that he is the head of household, he has the power to make all the decisions.
>
> (Ms Raksa, married, 21 years old, in-depth interview, 2018; Group 1)

Gender norms were enforced by how the husbands limited their wives' options to engage in income-generating activities outside fishing:

> I have noticed that the fish resources have declined over the last few years. Income from fishing is less than it was earlier, and I have to fish far into the lake to maintain my fish catches and increase my income to support the household demands. I am the head of my household and the main income earner. As long as I can maintain this household income, I won't let my wife work outside the household because no one can replace her with care for the children.
>
> (Mr Naeth, married, 45 years old, in-depth interview, 2018; Group 1)

The men in Group 2 were more flexible than the men in Group 1 about the gendered division of labour, and the women in Group 2 worked as daily wage labourers to earn independent incomes. Importantly, there were more female members in the families in Group 2, and household chores were done by other women, such as mothers-in-law, which meant that the wives were free to do other jobs:

> Income from fishing has decreased over the last ten years. My husband does not go fishing as frequently as he did earlier. He does not go fishing when the weather is not favourable, and he may look for occasional wage work in the village, since we cannot migrate outside the village. I mostly cut fish heads to earn some additional income, while my mother-in-law takes care of my small baby. I know the income I earn is not much, but I feel happy that I can help my husband to support the family and I can use my income to spend on what I want. It means that I am no longer dependent on his income only.
>
> (Ms Thida, married, 19 years, in-depth interview, 2018; Group 2)

Households in Group 2 still had a strong notion that women should do household chores and men should earn income. However, if there was more than one woman in the household, one of them could take care of the reproductive work while the other woman or women earned some income. The men's identities were closely linked to fishing, and therefore, all forms of livelihoods outside fishing were considered secondary. Similarly, women's incomes were considered secondary, and women could return to reproductive work whenever required. Thus, even though they earned an independent income, women's roles and statuses in their family did not seem to have changed:

> In my family, my brother and my father go fishing, while my mother is responsible for household tasks. I rarely do household work, even though I am a woman, because I need to earn money to support the household. However, sometimes I help my mother with household tasks when she is ill.
>
> (Ms Sreymom, single, 23 years, in-depth interview 2018; Group 2)

By contrast, households in Group 3 had stopped fishing altogether. Among those households, women had some experience of business, from which they were able to earn an income to support their family. The men in the same households did not oppose the women's income-earning activities and they too did some reproductive work to support the women's efforts to earn some income:

> Since my husband's income from fishing has decreased, our household income decreased. I talked to him and decided to ask him to stop fishing, since I saw that his health was not good enough to continue fishing. I decided to run a small retail business in the village to earn money for our survival. My husband used to disagree with my idea, but since he could no longer support the household, he let me try my business. I go to sell food by boat every day and can earn around 20,000 to 30,000 riel [USD5–7.5]. It is not much money, but it is enough to manage our daily expenditure. My husband does small-scale aquaculture at home and helps to look after the grandchild. I feel more tired than I did earlier, but I am happy that I can earn some money to help not only my household, but also myself.
>
> (Ms Panya, married, 58 years, in-depth interview, 2018; Group 3)

Some of the men in Group 3 had been persuaded by the women to stop fishing and find other income-generating activities:

> My husband decided to stop fishing because the income from fishing decreased and his health is not good. I learned to cook noodle soup from my mother. So, I decided to make a noodle soup to sell in the village. I convinced him and we decided to start this business after quitting fishing.
>
> (Ms Sarom, noodle seller, married, 46 years, in-depth interview, 2018; Group 3)

### *Summary*

From a comparison of attitudes and ideas across the three groups of households, we found that the perception of fishing as a man's occupation and the most important type of livelihood was typical in Group 1, but importantly, it was also a barrier that limited households from adapting to new livelihoods following the depletion of fish resources. Households in Group 2 were able to diversify because they had multiple women to manage household chores, and they had more members of working age. By contrast, households in Group 3 made the decision to move out of fishing and take up other occupations, and as a result, they were able to earn higher incomes than those in the other two groups. It was also noted that households in Group 3 were more flexible in terms of the gender divisions of labour and that men did reproductive work when required.

## Conclusions

Ethnic Vietnamese living in floating houses in Chhnok Tru village on Tonle Sap lake depend on the lake both as a place to live and for their livelihoods, because they are deprived of mobility and access to both land and other resources as a result of their liminal citizenship. They are not able to migrate to work, to obtain land for farming, or to find work in formal employment, such as in factories. Faced with decreasing fish resources, they needed to adapt their livelihoods. In this chapter, we have analysed the adaptation strategies of ethnic Vietnamese on Tonle Sap under conditions of restricted mobility. Despite the extreme limitations, there is still some flexibility in the different strategies that they employ: some continue fishing as before; some continue fishing but combine it with some small, secondary income-generating activities; and some stop fishing entirely and engage in other income-generating activities in the community. We found that the latter strategy provided the highest incomes for the households. Those who were able to move out of a livelihood dependent on fishing were able to adapt better to the diminishing fishing resources. We found that those who were able to go through the livelihood transformation had comparatively larger families, wider networks, and more knowledge and information about how to find new livelihood options, and importantly, they were more flexible in changing gender roles and expectations. In households in Group 3, which had transformed their livelihoods more than had households in Group 1 and Group 2, women had more say and more leadership. Unlike the households in Group 1, in which men resisted abandoning fishing, men in the households in Group 3 were more open to exploring other roles for themselves and even to take responsibility for childcare when women worked outside the home.

Women's empowerment and ability to transform gendered divisions of labour and norms are important resources (Kabeer, 2012; World Bank,

2011). While some of the ethnic Vietnamese fishers in the study village had limited resources and therefore difficulty in coping with the decreases in fishing resources, others had become more open to changing gender roles and that had allowed their households to adapt better. Although they had other networks and other forms of knowledge, in the absence of such changes in gender roles, they were not able to adapt to the reduced resources. In conclusion, for ethnic Vietnamese fishers who faced restricted access to resources, including as a result of limited motility, flexible gender relations are an important resource, which they can mobilize to help themselves to adapt to deceases in fish resources.

## Notes

1 According to Chapter 2, Article 6, of the sub-decree on community fisheries (Kingdom of Cambodia, 2005),

> A community fisheries . . . is a group of physical persons holding Khmer citizenship who live in or near the fishing area, voluntarily established and taking the initiative to improve their own standard of living by using and processing fisheries resources sustainably to contribute to economic and social improvement and poverty alleviation.

Community fisheries need to be approved by the Fisheries Administration (in the Ministry of Agriculture, Forestry and Fisheries) in order to be recognized as a CF.

2 Fish collectors buy fish directly from fishers and then sell the fish to fish wholesalers or other intermediaries.

3 Small fish are processed by cutting off the heads and removing the internal organs so that the fish can be used to make *prahok* (fermented fish). Cutting off the heads is the most laborious part of the process of making *prahok* and is usually done by women as daily wage labour.

4 The names of the participants are pseudonyms.

## References

Afifi, T., Milan, A., Etzold, B., Schraven, B., Rademacher-Schulz, C., Sakdapolrak, P., . . . Warner, K. (2016). Human mobility in response to rainfall variability: Opportunities for migration as a successful adaptation strategy in eight case studies. *Migration and Development*, 5(2), 264–274.

Amer, R. (2013). Domestic political change and ethnic minorities – A case study of the ethnic Vietnamese in Cambodia. *Asia-Pacific Social Science Review*, 13(2). Retrieved from https://ejournals.ph/article.php?id=9275

Ang, C., Weill, N., and Chan, J. (2014). *Limbo on Earth: An investigative report on the current living conditions and legal status of ethnic Vietnamese in Cambodia*. Minority Rights Organization, Investigative Research Report 2014, No. 2. Retrieved from http://mirocambodia.org/download/MIRO%27s%20Stateless%20Vietnamese%20Report%202014.pdf accessed 5/2/2020.

Bettini, G., and Gioli, G. (2016). Waltz with development: Insights on the developmentalization of climate-induced migration. *Migration and Development*, 5(2), 171–189.

Black, R., Bennett, S.R.G., Thomas, S.M., and Beddington, J.R. (2011). Migration as adaptation. *Nature*, *478*, 447–449.

Chandler, D. (2008). *A history of Cambodia* (4th ed.). Boulder, CO: Westview Press.

Ehrentraut, S. (2011). Perpetually temporary: Citizenship and ethnic Vietnamese in Cambodia. *Ethnic and Racial Studies*, *34*(5), 779–798.

FAO. (2013). *Mainstreaming gender in fisheries and aquaculture: A stock-taking and planning exercise. Final report*. Rome: FAO.

Ha, V.S. (2008). The harmony of family and the silence of women: Sexual attitudes and practices among rural married women in northern Viet Nam. *Culture, Health & Sexuality*, *10*(S1), S163–S176.

Haque, M.M., and Kusakabe, K. (2005). Retrenched men workers in Bangladesh: A crisis of masculinities? *Gender, Technology and Development*, *9*(2), 185–208.

Jeanne, E., Sophy, E., and Sareivouth, T. (2013). *Early childhood development education for floating villages project*. Project ID: P146085, Final report. Retrieved from https://docplayer.net/81206136-Early-childhood-development-education-for-floating-villages-project-project-id-p146085.html accessed 5/2/2020.

Kabeer, N. (2012). *Women's economic empowerment and inclusive growth: Labour markets and enterprise development*. SIG Working Paper 2012/1. Retrieved from www.idrc.ca/sites/default/files/sp/Documents%20EN/NK-WEE-Concept-Paper.pdf

Kaufmann, V. (2002). *Re-thinking mobility: Contemporary sociology*. Aldershot: Ashgate.

Kiernan, B. (1990). The survival of Cambodia's ethnic minorities. *Cultural Survival Quarterly Magazine*, September. Retrieved from www.culturalsurvival.org/publications/cultural-survival-quarterly/survival-cambodias-ethnic-minorities accessed 5/2/2020.

Kingdom of Cambodia. (2005). *Sub-decree on Community Fisheries Management*. Subdecree No. 80/OrNorKror/BorKor. Retrieved from www.ecolex.org/details/legislation/sub-decree-no-80ornorkrorborkor-on-community-fisheries-management-lex-faoc085864/ accessed 5/2/2020.

Kirchner, L.M. (2015). *Living on the margins: On the status and standing of minorities and indigenous peoples in Cambodia*. Retrieved from https://kh.boell.org/sites/default/files/hbs_living-on-the-margins_a5_3_rz-online-with-publishing-date.pdf accessed 5/2/2020.

Kong, M., and Nachemson, A. (2017). I have no feeling for Vietnam. I only live in Cambodia. *Phnom Penh Post*, 28 November. Retrieved from www.phnompenhpost.com/national-post-depth/ethnic-vietnamese-some-living-cambodia-generations-see-documents-revoked accessed 5/2/2020.

Kymlicka, W. (1995). *Multicultural citizenship: A liberal theory of minority rights*. Oxford: Oxford University Press.

LaBorde, P. (1996). *Vietnamese cultural profile*. Reviewed and updated December 2010 by Oanh Doan and S. Gruen. Retrieved from https://ethnomed.org/culture/vietnamese/vietnamese-cultural-profile accessed 5/2/2020.

Long, L.D., Hung, Le Ngoc, Truitt, A., Mai, Le Thi Phuong, Nguyen Anh, Dang. (2000). *Changing gender relations in Vietnam's post Doi Moi era*. Policy Research Report on Gender and Development, Working Paper Series No. 14. Retrieved from http://documents.worldbank.org/curated/en/564931468315308766/pdf/341320Gender0wp14.pdf accessed 5/2/2020.

Minority Rights Organization. (2015). *Research situation regarding legal document of Khmer Khrom*. Retrieved from http://mirocambodia.org/?p=692 accessed 5/2/2020.

Minority Rights Organization. (2017). *Decreasing statelessness and raising legal awareness among former Vietnamese and Khmer refugees in Cambodia*. Retrieved from http://mirocambodia.org/?page_id=328

Nguyen, L., and Sperfeldt, C. (2012). *A boat without anchors: A report on the legal status of ethnic Vietnamese minority populations in Cambodia under domestic and international laws governing nationality and statelessness*. Jesuit Refugee Service ((JRS). Retrieved from https://jrscambodia.org/publication/boat_without_anchors_jrs.pdf JRS Cambodia

Parsons, L., and Lawreniuk, S. (2018). Seeing like the stateless: Documentation and the mobilities of liminal citizenship in Cambodia. *Political Geography*, *62*, 1–11.

Phanith, C. (2013). The implications of hydrology change on local people's livelihoods around Tonle Sap Lake: A case study in Chhnok Tru, Kampong Chhnang Province in Cambodia. *Hydrology Change Research Paper*, *1*, 97–116. Retrieved from www.researchgate.net/publication/264539913_The_Implications_of_Hydrology_Changes_on_Local_People's_Livelihoods_around_Tonle_Sap_Lake_A_case_study_in_Chhnok_Tru_Kampong_Chhnang_Province_in_Cambodia_Chou_Phanith accessed 5/2/2020.

Serrat, O. (2005). *Threats to the Tonle Sap,* Asian Development Bank.

Shelton, B.A., and John, D. (1996). The division of household labor. *Annual Review of Sociology*, *22*(1), 299–322.

Sperfeldt, C. (2017). *Report on citizenship law: Cambodia*. GLOBALCIT Country Report 2017/02. Retrieved from https://cadmus.eui.eu/handle/1814/45084 accessed 5/2/2020.

Takizawa, S., Kurisu, F., and Satoh, H. (2009). *Southeast Asian water environment 3*. IWA. Retrieved from www.iwapublishing.com/books/9781843392767/southeast-asian-water-environment-3 accessed 5/2/2020.

World Bank. (2011). *World development report 2012: Gender equality and development*. Washington, DC: World Bank.

# 9 Mobility in contexts of precarity

## Kin solidarity and migrant networks among small-scale fishers in coastal Tamil Nadu

*Nitya Rao, R. Manimohan and C.M. Pratheepa*

### Introduction

Migration experiences are not uniform, and in this chapter, we demonstrate how different types of migrants from varied physical locations and social contexts both access and use kinship and social networks differently, which in turn influences their migration experiences, patterns, and well-being outcomes. The reasons for migration vary, but they are related to resource depletion, growing environmental uncertainty, and climate-related stress, with changes in rainfall patterns, for example, having a negative impact on coastal breeding habitats (such as those of mangroves) and on spawning seasonality (Vivekanandan, 2011), alongside changes in opportunities for employment in agriculture. Both types of impacts potentially contribute to the precarity of labour relations, in that men are pushed into migration and often employ risky mobility strategies to fulfil their role as providers for their households. While there is growing evidence that links social capital to resilience to climate stress (Goulden et al., 2013; Jordan, 2015), power relations are central to any discussion of social capital or social networks, because obligations and commitments, in addition to claims and entitlements – all of which are located in social and gender ideologies – can be supportive in some cases but limiting in others.

This chapter is based on survey data and qualitative research conducted over the course of one year, from March 2017 to February 2018, when we carried out fieldwork in two districts of the southern Indian state of Tamil Nadu: Kanyakumari District on the west coast, which is dominated by the Roman Catholic Mukkuvar caste, and Cuddalore District on the east coast, where the majority of the fishers belong to the Hindu Pattinavar caste. We unpack the differences in the quality of social relationships and networks across different migrant streams in the two locations: migration for multi-day fishing, migration overseas for fishing and other types of work, and distress migration. We conducted a survey of 300 households in each district and held in-depth interviews with women and men selected from different categories of fishers, as well as key informant traders, auctioneers, labour recruiters, and village leaders.

In the next section, we briefly discuss the key conceptual ideas that informed our analysis of social and gender relations in contexts of migration in the two study districts, and thereafter, we discuss in detail how they work in practice across the different migration streams in those districts. Whereas fishing is a caste-based occupation in Tamil Nadu, and caste itself forms a basis for both inclusion and exclusion, the support provided by the Church in Kanyakumari District, albeit gendered, appears to provide a depth of trust and reciprocity in social relations that is not visible in Cuddalore District.

## Social and gender relations in migration

There is considerable debate on a precise definition of *social capital*, but the term is generally understood to refer to actors' ability to secure benefits through membership and participation in social networks or other social structures (Bourdieu, 1977; Portes, 1998). It refers to aspects of social relationships that may be converted into other forms of capital (e.g. economic) and ultimately help actors to maintain or improve their position in society (Bourdieu, 1977; Portes, 1998). Although social capital reflects a somewhat utilitarian perspective, our interest here is in exploring different types of social relationships and networks, including those based on kinship, shared identity (e.g. religion, ethnicity), shared location or shared history, or professional arrangements. These have also been referred to as bonding and bridging relationships (Woolcock and Narayan, 2000) (for further discussion, see Chapter 1, this volume). The ties that bind these relationships may be strong or weak (Granovetter, 1973), and although they provide some backup for unemployment or a collapse in income, they are not free of intergenerational and gendered conflicts and hierarchies (Silvey and Elmhirst, 2003). The potential capacity to construct purpose, meaning, and utility from social relationships can be restricted by inequitable social and economic structures and institutions that in turn reflect social values and norms (Cleaver, 2005; Molyneux, 2002). This is visible particularly in the case of migrant workers who enter marine fisheries as crew members in the study localities (see Chapter 2, this volume).

In the Indian context, caste shapes social relations among small-scale fishers. The main fishing caste in Cuddalore is the Pattinavars, and the main fishing caste in Kanyakumari is the Mukkuvars. Members of both castes own the boats and fishing gear, and they pride themselves on having the required skills to fish at sea. Caste affinity is clearly a form of support, but when constructed in masculine terms, it can also impose obligations that are likely to disadvantage those who are not in leadership positions or who are not members of the same caste, which in the study districts are migrant labourers. Support can involve information about labour opportunities or access to markets. However, benefiting from such support can incur costs (e.g. in the form of labour or favours), which if not met can mean individuals are not seen as 'doing the right thing'. The task of managing such tensions is left to close family members, usually wives, who provide access to close kin and neighbourhood relationships, but equally smooth over any potential tensions by ensuring degrees of reciprocity and mutual help are maintained. If limited resources cease to be available or become diluted – for example, due to men's alcoholism – it can increase conflict, tension, and strain in household relationships (Cleaver, 2005; Shelton, 2017). In such cases in Kanyakumari, the role of the Church becomes important because it provides an additional layer of support to both men and women by encouraging reciprocity and mutuality in everyday relations.

Furthermore, the type and nature of social relations, whether informal or formal, helps to explain particular migration patterns (Palloni et al., 2001). Informal migrant networks are the interpersonal ties that connect migrants and non-migrants, both in the place of origin and at the destination, through kinship, friendship, and shared community origin. Such networks provide both direct assistance and access to information from prior migrants, thereby reducing the costs (economic, social, and psychological) for potential migrants (Garip, 2008; Massey and Garcia-España, 1987). Such networks and relationships, which are particularly visible among the Kanyakumari fishers, are negotiated and rely on principles of trust and expectations of reciprocity. Hence, the poor who have limited network connections may be constrained in their mobility strategies (Cleaver, 2005; Shelton, 2017; Silvey and Elmhirst, 2003; Woolcock and Narayan, 2000).

Formal networks can be referred to as the 'migration industry', and they comprise travel agents, brokers, housing agents, recruitment agencies, traffickers, and others who have an interest in and continue to facilitate migration (de Haas, 2010). Such formal networks are required for Indian labourers who migrate to the Gulf countries and elsewhere (e.g. South East Asia, Singapore, and Malaysia) and can be seen in operation in Cuddalore District. Whereas attention to the 'migration industry' is often focused on its negative aspects related to exploitation, especially in relation to poorer or lower-skilled migrants, Baas (2018) highlights that the industry also provides access to migration that would otherwise be out of their reach. Nevertheless, significant information asymmetries remain between brokers and labourers, often resulting in 'failed' migration, and are reflected either in incomes that are not remunerative or in attempts to circumvent the system by working illegally. The initial disadvantages experienced by such migrants, whose debts are often financed by their families and kinship networks, may have an impact on their migration outcomes by reinforcing income inequalities and relative deprivation (de Haas, 2010).

Social networks are also gendered. Typically, women are not members of networks that bring economic advantages or operate in masculine spaces, which in the study districts is marine fishing. Rather, women's networks tend to rely more heavily on non-monetized labour and time exchanges that can be accommodated within domestic labour divisions (Molyneux, 2002;

Smith, 2000). Data from a range of countries demonstrate that women in low-income groups often have stronger community and kinship ties and engage in supportive and reciprocal relationships, such as the Church or other local association activities (Molyneux, 2002), and are not restricted to their reproductive functions. In the study context, especially in Kanyakumari District, women engage in a range of social networks in order to manage cash flows, both for business (investment in fishing operations) and to maintain their households (Ram, 1991), especially in the absence of the men in their households. Despite carrying the burden of financial management, they are compensated by engaging in and leading political and social activities through their church-organized neighbourhood groups (*anbiams*). Such support and such organization are lacking in Cuddalore District, thus making the masculinization of fisheries and the exclusion of women more starkly visible. The women's participation in reciprocal kin networks comes at high costs; for example, the pressure to make substantial contributions to the payment of dowries has led to negative implications for their health and well-being, increased their work obligations, and threatened the survival of girls (for additional information, see Chapter 2, this volume).

## Migration context in coastal Tamil Nadu

As noted in Chapters 2 and 7, the increase in the numbers, motorization, and density of fishing vessels in Tamil Nadu since the 1960s and more rapidly since the 2004 tsunami has led to increased competition among the fishers and to decreased catches per unit (Bhathal, 2014, pp. 98–99). To compensate, fishers and fishing communities have engaged in various strategies to maintain viable fishing livelihoods, included fishing farther away and migrating (Bavinck, 2008; Rajan, 2002).[1] Social relations and networks contribute differently to the adoption of particular strategies and outcomes according to their access to resources and assets, especially fishing vessels and fishing gear.

In the case of coastal Tamil Nadu, migration patterns internally and overseas have varied widely depending on the length of time away, distance from the place of origin, and the purpose of the movement, and boundaries have not always been clear. In recent years, four distinct migration streams have emerged: overseas for fishing, overseas for non-fishing activities, multiday fishing, and distress migration (Table 9.1). As shown in Table 9.1, the majority of migrants from both districts engaged in multiday fishing in 2017.

The reasons for particular migration choices have varied, as shown in Table 9.2. A Mann-Whitney test was performed to test for significant differences in reasons for migration between the two districts. Of the 16 reasons for migration, 10 were statistically significant. While improving the family's financial situation was common in both of the study districts, financing marriages, education, or healthcare were important in Kanyakumari

*Table 9.1* Migration destinations for migrants from Cuddalore District and Kanyakumari District, Tamil Nadu, 2017

| *Migration destination* | *Cuddalore District* | | *Kanyakumari District* | | *Total* | |
|---|---|---|---|---|---|---|
| | *No.* | *%* | *No.* | *%* | *No.* | *%* |
| Samc district | 0 | 0.00 | 7 | 2.12 | 7 | 1.09 |
| Another district | 27 | 8.57 | 8 | 2.42 | 35 | 5.43 |
| Kerala or another state | 1 | 0.32 | 1 | 0.30 | 2 | 0.31 |
| Gulf countries | 15 | 4.76 | 44 | 13.33 | 59 | 9.15 |
| Singapore or Malaysia | 23 | 7.30 | 0 | 0.00 | 23 | 3.57 |
| Another country | 3 | 0.95 | 1 | 0.30 | 4 | 0.62 |
| Multiday fishing grounds | 246 | 78.10 | 269 | 81.52 | 515 | 79.84 |
| Totals | 315 | 100.00 | 330 | 100.00 | 645 | 100.00 |

Source: Household Survey 2017

*Table 9.2* Reasons for migration from Kanyakumari District and Cuddalore District, Tamil Nadu, 2017

| *Reason for migration* | *District* | | | | *Total no. of people* | *Statistically significant difference* |
|---|---|---|---|---|---|---|
| | *Kanyakumari* | | *Cuddalore* | | | |
| | *No. of people* | *%* | *No. of people* | *%* | | *p-value* |
| To earn money to construct a house | 40 | 40 | 60 | 60 | 100 | 0.003* |
| To earn money to buy a share in a ring seine unit | 0 | 0 | 42 | 100 | 42 | 0.00** |
| To earn money to buy a share in a mechanized boat | 1 | 20 | 4 | 80 | 5 | 0.187 |
| To earn money for marriages | 132 | 81 | 30 | 19 | 162 | 0.00** |
| To finance children's education | 166 | 80 | 41 | 20 | 207 | 0.00** |
| To finance family health expenses | 152 | 96 | 7 | 4 | 159 | 0.00** |
| To improve the family's overall economic condition | 294 | 54 | 253 | 46 | 547 | 1.00 |
| Local conflicts | 1 | 50 | 1 | 50 | 2 | 1.00 |
| To repay debt | 44 | 72 | 17 | 28 | 61 | 0.003* |
| Depletion of resources at the home fishing ground | 10 | 8 | 113 | 92 | 123 | 0.00** |
| Better fishing opportunities for own fishing unit at the destination | 12 | 60 | 8 | 40 | 20 | 0.651 |

(*Continued*)

*Table 9.2* (Continued)

| *Reason for migration* | *District* | | | | *Total no. of people* | *Statistically significant difference* |
|---|---|---|---|---|---|---|
| | *Kanyakumari* | | *Cuddalore* | | | |
| | *No. of people* | *%* | *No. of people* | *%* | | *p-value* |
| Better resources and/or fishing potential at the destination | 44 | 32 | 94 | 68 | 138 | 0.00** |
| Better infrastructure and facilities at the destination | 5 | 100 | 0 | 0 | 5 | 0.128 |
| Better wages and/or shares at the destination | 7 | 26 | 20 | 74 | 27 | 0.003* |
| Following earlier migrants who left for better job opportunities outside fisheries at the destination | 0 | 0 | 18 | 100 | 18 | 0.001** |
| Shortages of skilled male fishers at the destination | 3 | 100 | 0 | 0 | 3 | 0.253 |
| Total no. of people in surveyed households | 911 | – | 708 | – | 1619 | – |

Notes: * significance level $p < 0.05$; ** significance level $p < 0.001$

Source: Household Survey 2017

District, and the depletion of fishing resources, house construction, and the purchase of boats (or shares in boats) were important in Cuddalore District.

In the next section, we discuss each of the four migration streams, unpack the nature of ties on which the migrants rely, and discuss the implications for their success.

## Social networks and relations in different migration streams

### *Multiday fishing*

As shown in Table 9.1, multiday fishing is by far the most important pattern of movement in both of the study districts. The multiday trips typically range in length from three days to seven days or a fortnight in the case of Cuddalore District, whereas in the case of Kanyakumari District, the fishing trips last longer, often a month or more, because the fishing vessels are frequently moored at larger ports (and not always owned by someone from the village) and the fishers who work on them stay at the ports between trips.

In Kanyakumari, the crew generally comprise 11–12 members, with the captain responsible for sourcing crew members. Given their greater responsibility, the captains also earn more in both money and prestige and can ensure that the crew receive fair remuneration, whether as a share of the catch or as a wage. Crew members are often family members or friends, with whom there is already a relationship of trust, although there may be some 'outsiders' (e.g. in-migrants from northern states in India). The study participants reported that working with family members was more manageable than working with outsiders because the former were aware of the investments made by their families and were willing to work through any disagreements that arose on the vessel: 'To make a living, they have to adjust among themselves. They know that the investment is heavy and they have to yield to situations to keep things smooth' (Leonard, from a village in Kanyakumari District).[2] Given the high levels of risk at sea (see Chapter 7, this volume), the limited access to information, and the reality that fishing is the only option available to them to earn enough money for their needs (Bavinck, 2014), those onboard must work together and support each other. In relying on strong ties to work cooperatively, most often in bonding relationships, the fishers' practices also demonstrate their ability to build trust and work with 'outsiders' on the vessels.

In the multiday fishing households, women play important roles in organizing and participating in social events, such as marriages and funerals. They are members of the neighbourhood *anbiams* and thus are responsible for the management and governance of the household and community. The women also engage in regular collective prayers for the well-being of the male fishers. While the women having considerable power in the public domain, the Church presents their activities as the performance of their roles of responsible mothers and wives, who support their husbands. The Church's perspective was thus an example of how ideas that link motherhood to women's roles in maintaining social relations and a sense of community are reinforced (Molyneux, 2002).

In Cuddalore District, where multiday fishing is for relatively shorter durations than in Kanyakumari District, the boats return to the harbours closest to their homes so that the men can visit their families. In this case, women in Cuddalore District are less engaged in the fishing enterprises and have less control over the income from them, despite contributing to raising money for fishing gear or fishing trips. New groups of men who owned shares in boats, supported by traders and exporters, are responsible for the management of much of the money (see the subsection 'Capital mobilization' in Chapter 2, this volume). The social networks relating to fishing are increasingly masculinized and formalized, while women lose some of the power they once held, when fishing was a small-scale family enterprise and boats landed on the village beach. The new forms of boat ownership represent a shift from a focus on bonding relationships based on strong family ties (between fathers and sons or between brothers) to relationships based on shared identity as fishers, often entailing weaker ties, as well as bridging relationships (between different occupational groups) consisting of a mix of strong and weak ties.

### *Overseas for fishing and other employment*

A number of men from Kanyakumari District migrated to the Gulf countries (e.g. Qatar, Bahrain, and Saudi Arabia) to fish there (Table 9.1). They are known for their fishing skills and are therefore prized as workers on fishing vessels (see Chapter 7, this volume). They fish in the Gulf for nine months per year and return home between November and February, for Christmas. The migration stream seems to have begun in the mid 1990s, when a small number of men went to work as crew on fishing vessels in the Gulf countries. Gradually, they earned the trust of boat owners from the Gulf countries through hard work and earned the opportunity to rent the fishing vessels and serve as captains. The men then started to provide opportunities for other men from their villages (either family or friends) to travel to the Gulf and work on fishing vessels, each of which required 15–20 people as crew. Currently, fishers cover the cost of fishing gear and equipment (e.g. GPS, fuel, hooks, and lines) and their board and accommodation. They view their migration as an investment in their families through funding their children's education, a more comfortable lifestyle, and better marriage prospects for their children. Access to the jobs depends on bonding social capital and close ties, which are based on shared identity (e.g. family or village) and trust between individuals.

While the male fishers provide information and connections, the boat owners have to sponsor the visas of their crew. Recent changes to Saudi laws have resulted in a reduction in the number of individuals that a Saudi boat owner is able to sponsor without paying additional costs to obtain visas for the migrants; therefore, the migrants are often willing to cover these costs. Fishing in the Gulf is tightly regulated, and the migrants must sign in and sign out when they leave the harbour. Thereafter, they must sign in at the harbour at least once every seven days. This enables the harbour police to keep track of the migrant fishers, which limits the amount of time that they can spend at sea. Despite this, fishing in the Gulf is more lucrative, and the migrant fishers' income is higher than that of fishers who stay in their home villages.

Some educated young men (ages 18–35) have been successful in finding white-collar jobs in the Gulf or in other parts of India. As in the case of fishing, the primary route for securing the jobs has been through informal networks, such as a family member abroad, rather than through agents. Cousins, sons, and sons-in-law help each other find work, thus demonstrating the importance of bonding social capital with strong ties to access more lucrative employment. With four daughters, Denzil in Kadalkarai Uuru, Kanyakumari District, had incurred considerable debt in connection with the marriage of his daughters. With the help of his son-in-law, his son had secured a job as an electrician in the Gulf and was gradually repaying the loans.

With the men away for nine months of the year, the women are left responsible for all household matters, such as financial arrangements for everyday expenses, and for social and ritual events. Although they often initially turn to their kin networks when they want support, the resources available in the relationships are insufficient to meet household needs. As noted in the introduction to this chapter, the women seek support from the *anbiams*, which includes helping them with their savings, yet often the financial capital available from the groups too is insufficient. Women then take out *kanthu vatti* (loans from private lenders at high interest rates) to meet their short-term needs. Thus, the men's migration has led to a strengthening of women's networks in the public space, which have been facilitated both by the Church and through their access to and management of multiple sources of money (for additional information on the role of the Church in supporting women when the men are away from home, see Chapter 7, this volume).

In Cuddalore District, young men in their twenties, many of whom are educated with engineering degrees and looking for either skilled jobs or white-collar employment, migrate to Southeast Asia, and comparatively fewer migrate to the Gulf countries. They have to complete a three-month training course and take a test to obtain visas to move to Singapore or Malaysia. The process is complex, and although they follow friends or relatives, their ties are weaker and they need the help of agents. The young men have to bear all of the costs of migration and often do not secure the jobs or incomes that the agents promised them would match their qualifications. Many return home disappointed, so they increasingly view this type of migration stream as a short-term strategy to raise financial capital to purchase shares in a boat or to raise money for dowry payments, in addition to meeting consumption expenses. Dowries are a way of raising capital for buying shares in boats, and therefore, brothers or fathers migrate to raise money for dowry payments for their sisters or daughters respectively. While dowries do not contribute directly to the men's own well-being, the practice is seen as a responsibility that contributes to the potential well-being of sisters and daughters and thus to the men's respectability as responsible men. For such men, fishing locally, especially if they own a share in a boat, is more lucrative than daily wage work overseas. Arul's second son had started fishing with his father after he was unable to find suitable work abroad, and his father said, 'It is a great disappointment that even engineering graduates get a meagre pay as low as 10,000 rupees [USD141]. He is now fishing with me.'

### *Distress migration*

People from the northern states of India (e.g. Uttar Pradesh, Assam, and Bihar) have moved to some coastal areas of Tamil Nadu in response to drought or lack of income-earning opportunities in their home locations. In-migrants move to earn money and are willing to work hard. They start work as on-shore labourers, assisting with loading and unloading vessels (e.g. fishing gear, ice, and fish), crushing ice, and mending nets, and then work their way up to become cooks and then crew members on the boats. The ice loaders and ice crushers work 10 months per year and then return to their place of origin during the seasonal ban on fishing for large boats, which corresponds to the fish-breeding season. While most of the in-migrants hear about the work opportunities through their family networks, on their arrival in the districts, they access jobs through labour contractors.

In Kanyakumari District, where the crews generally number 11–12 per boat, only two or three may be migrants, and the other crew members come from fishing households, a key reason being the need for crew with 'skills' who are able to live at sea for 15–30 days and weeks at a stretch and make responsible decisions. In many ways, the in-migrants remain anonymous, demonstrating the potential for exclusion in bonding social relationships (i.e. with those who are not migrants). This exclusion was evident in December 2017, when a severe cyclone affected Kanyakumari District, especially the Thoothoor cluster of villages, resulting in over 100 fatalities (108 fishers according the Church records). The Church provided support and negotiated for compensation on behalf of the Mukkuvar fishers who had lost their lives or incurred losses, but no representative or group acted on behalf of the in-migrants. They could not even be identified by name and did not receive any compensation. Disasters (including those linked to climate stress, such as cyclones) are often framed as opportunities for positive social change, yet the events can also present opportunities for elites to use exclusionary forms of social capital to perpetuate inequality or to solidify existing power structures (Jordan, 2015).

The majority of the in-migrants were men who had travelled alone, but a small proportion were accompanied by their wives. Many lived in dilapidated housing that had been abandoned after the 2004 tsunami, because they were the only places they could afford to rent. With limited assets locally and without social assets in terms of family or religious groups to sustain them, they had to rely on their work and income; in the absence of work and income, they had to return to their home villages. Despite the difficulties they faced, their numbers had increased rapidly between 2012 and 2017.

In Cuddalore District, each ring seine fishing unit requires 30–40 crew members. Such numbers are not available in the local community, and therefore, distress migrants from elsewhere in the district, who have no work once the rain-fed crops have been planted, may be recruited by boat owners to work as casual labourers on their boats. Sometimes up to half of the entire crew may come from farming villages. However, the crew members may also come from farther afield (e.g. northern states of India). Those without fishing experience find the work through informal networks of relatives and friends rather than through formal means, as Kaiser told us:

> I came to the locality 16 years ago as an ice cream seller. The business was good in the summer, but not winter. I had built contacts with the local fishers over many years, so tried my hand as a crew member one winter. This paid well, so several young boys from my village have joined me now.

All of the in-migrants are seasonal distress migrants, who plough their small plots of land at home and migrate to Cuddalore District to work on ring seine boats for the six-month fishing season. They are identified as employees on vessels but are paid lower wages than those who own shares in the boats and are unlikely to earn enough to purchase such shares or to become boat owners. While accepted as workers by their respective village fishers' society, boat ownership is restricted to the Pattinavar fishing caste in Cuddalore District. The in-migrants rely on bonding social capital to access employment opportunities on vessels, but the strong bonds of caste and kinship within the Pattinavar caste exclude them from accessing more lucrative opportunities that can be gained through boat ownership.

Another form of distress migration in Cuddalore District is the out-migration by the nontraditional lagoon fishers who belong to the scheduled tribe known as the Irulars to collect worms farther south on the Coromandel Coast. The live worms are used as feed on aquaculture farms.[3] Both men and women engage in the work. They lack alternative livelihoods or social networks. The work is organized through agents and subagents, who pay one-way transport costs for a husband-and-wife team and pay a weekly sum of INR100 (USD1.4) to cover groceries and 1 kg of rice, in addition to the piece rate payment for the worms collected. Sometimes the agents and subagents cheat the workers by under-weighing the worms or by reducing the price paid per kilogram. The work is physically hard and environmentally destructive, which brings the Irulars into conflict with the traditional fishers. However, due to a lack of rain and shortages of fish in the lagoon, they have no other option than to engage in the worm-collecting activity. Socially, they are constructed as a class of labouring poor with few bridging relationships or linking ties, who therefore remain tied to oppressive labour relations.

## The social landscape, gender relations, and well-being outcomes

The discussion in the preceding section ('Social networks and relations in different migration streams') points to the various migration pathways in the study districts and to the nature of social relations and networks, formal and informal, bonding or bridging, which facilitate or hinder the movements. There are differences between the two districts. Clearly, there is a stronger network of ties in Kanyakumari District than in Cuddalore District, both within kin groups and in the community through the Church-led *anbiams*. This insight into the strength of social networks as perceived by women and men in the two study districts is evident from Table 9.3, which lists the average of scores (range 1–5 on a Likert scale). In Kanyakumari, the role of the Church and neighbours is clearly significant, whereas in Cuddalore District, both relatives and friends and traders and agents appear to play more important roles. For men in Cuddalore, community activities appear crucial, reflecting both the partnerships in boat ownership and the membership of their village fishers' society, which provides subsidies for fuel and compensation during the fishing ban period. A greater dependence on building interpersonal relationships in Cuddalore has gender implications in that it contributes to a decline in women's status rather than a strengthening of their social position, the latter of which is the case in Kanyakumari.

The central role of social networks, apart from logistical support for new and potential migrants, relates to the process of mobilizing financial capital. In Kanyakumari District, fishers support the Church: they provide one-tenth of their catch from every fishing trip as an offering for God's protection during the trips. Additional donations are made for Church feasts, new building works, and membership expenses (which cover the running costs of the Church). Furthermore, the Church has helped fishers by supporting the establishment of fishers' cooperatives, thereby enhancing their control over prices and their ability to sell directly to wholesalers rather than through traders. While some cooperatives also run chit funds to provide financial support to members (typically running for approximately 3.5 years), much of the mobilization of financial capital is done by women from a variety of sources, including church-based savings groups, loans raised on jewellery (from banks, private banks, and pawnbrokers), and loans from friends. The networks, whether based on religion, kinship, or friendship, rely on their bonding social capital: on relations of reciprocity and trust. In Cuddalore District, women rely on loans from private moneylenders and traders and on those raised from their jewellery (Figure 9.1). The jewellery-based loans are central in women's domains and strongly associated with their social status. Jewellery is largely received by women from their natal kin as dowry during their marriage. They also tend to convert their income into jewellery, as it is seen as a more reliable form of savings than cash. Loans taken against their jewellery from banks or private companies are used to support investments in the purchase of boats or boat shares, which are men's assets. They are also used to finance lifestyles, children's education, and dowries, all of which contribute to the building of social capital, with the potential to pay economic returns (Bourdieu, 1977).

According to the study participants, there was dowry inflation in both Cuddalore District and Kanyakumari District. The costs are significant, and families need to sell property or take out loans to meet them. It is also costly

*Table 9.3* Strength of social networks perceived by women and men in Kanyakumari District and Cuddalore District, Tamil Nadu, 2017 (measured on a 1–5 Likert scale)

| *Statement* | *Women (mean value)* | | | *Men (mean value)* | | | *Total Kanyakumari District (mean value)* | *Total Cuddalore District (mean value)* |
|---|---|---|---|---|---|---|---|---|
| | *Kanyakumari District* | *Cuddalore District* | *Total women* | *Kanyakumari District* | *Cuddalore District* | *Total men* | | |
| I have relatives and/or friends in the village | 3.42 | 3.52 | 3.47 | 3.06 | 3.64 | 3.34 | 3.24 | 3.58 |
| I have relatives and/or friends in the district or in neighbouring districts | 2.96 | 3.44 | 3.20 | 2.83 | 3.42 | 3.12 | 2.895 | 3.43 |
| My neighbours are supportive of my family at times of crisis | 3.96 | 3.56 | 3.76 | 3.90 | 3.51 | 3.71 | 3.93 | 3.535 |
| The leaders in my religious community are actively supportive of the community | 4.23 | 2.89 | 3.56 | 4.02 | 3.01 | 3.52 | 4.125 | 2.95 |
| I have contact with several traders and agents who provide credit when needed | 2.11 | 2.61 | 2.36 | 1.96 | 2.95 | 2.45 | 2.035 | 2.78 |
| I regularly participate in community activities | 4.51 | 4.80 | 4.65 | 3.83 | 4.82 | 4.32 | 4.17 | 4.81 |

Source: Household Survey 2017

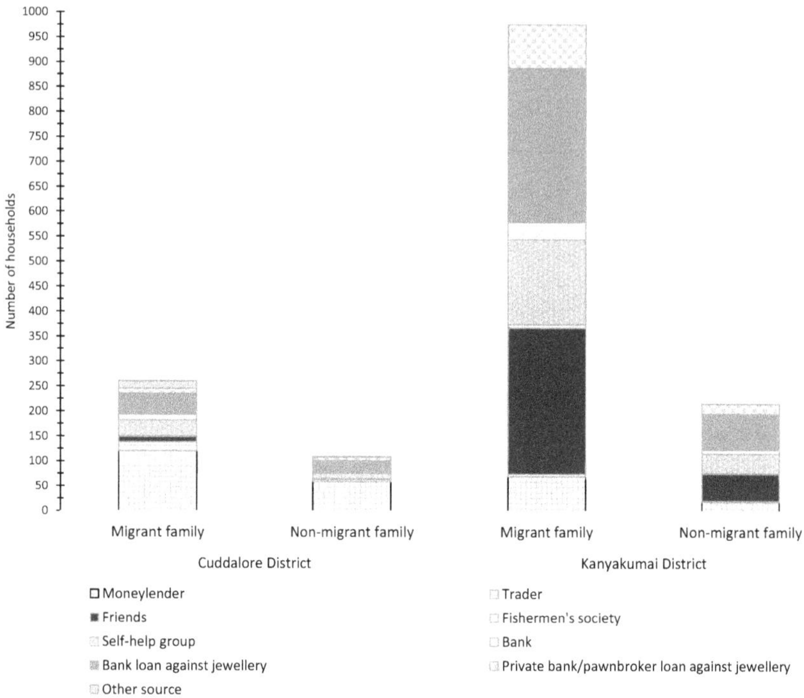


*Figure 9.1* Comparison of debt sources in Cuddalore District and Kanyakumari District, Tamil Nadu, 2017

Source: Household Survey 2017

to educate women, especially as they are unlikely to work and earn enough to cover the costs incurred. Nevertheless, their education is seen as important for their social reproduction roles, which includes money management and the supervision of the education of their children, especially when the men are absent. Sons are expected to contribute to their sisters' dowries, and they expect to recoup the expenses through the dowries that they receive when they marry. Many young men feel burdened by this social responsibility, yet there seem few opportunities to avoid it. A key reason for the increase in dowry is the need to demonstrate prosperity and thus maintain 'status' in the context of uncertain fish catches and incomes. However, the demonstration of prosperity that can be seen in the education of children, the building of expensive homes, the ownership of boats, or generous donations to the Church, temple, or other traditional institutions is financed by loans rather than by income. Those who have access to financial capital, such as boat owners or those who have multiple shares in boats, are thus seen as reliable (and therefore more likely to have credit extended to them or to be eligible to participate in chit funds), which enables them to enhance their social status further still. However, this can further exclude those without access to financial or social resources. Our survey data and qualitative research revealed that whereas there was more wealth in Kanyakumari District than in Cuddalore District, at least in external markers such as housing, education levels, water, and sanitation, a number of participants noted low levels of well-being and said that their wealth was not 'real', but rather, their income was used to service and repay debts (for a discussion on well-being, see Chapter 7, this volume).

## Conclusions

As a caste-based occupation in Tamil Nadu, fishers' primary social relations and networks are other fishers, namely friends and relatives who share a common occupation, skills, and, to some extent, aspirations. However, with rapid changes in technology and patterns of fishing, caste relations are no longer sufficient to secure livelihoods or fulfil fishers' aspirations. With men increasingly moving farther away from home for both fishing and non-fishing work, the success of their ventures is linked to the strength of the social ties and networks that they are able to build and access. Moreover, such ties and networks are gendered.

In cases when fishers' skills can be flaunted as a positive attribute, their fellow caste members can support the migration of their kin and peers by influencing employers to sponsor their migration, if not all of the costs. This is still the primary form of social networking in Kanyakumari District, where it is also supported by the Church. With more industrial, capital-driven forms of fishing emerging, recruitment is often undertaken by traders and agents and not necessarily by fishers themselves, as in the case of Cuddalore District. These networks are more formal, but the ties are also weaker, leading to precarity in outcomes for the migrants. This became particularly apparent after the 2017 cyclone, which led to the deaths of over 100 fishers, although the dead migrant workers could not be identified in the absence of a representative or group to act on their behalf.

Women's roles have diverged in the two study contexts. In Kanyakumari District, women remain in control of financial management to a large extent, whereas in Cuddalore District, women have been marginalized from the fishing enterprises. Although in both cases, we found that dowry had been inflated to finance boats and a range of status-producing activities, in Cuddalore District, it had negatively affected women's survival chances. Hence, although kin solidarity remains crucial to successful migration, it has negative effects on women's status and well-being in contexts of growing uncertainty and precarity.

The broader context of migration in Kanyakumari District and Cuddalore District is situated in resource depletion, growing environmental uncertainty, and climate-related stress. While social capital has been linked to resilience to climate stress, this chapter also demonstrates that while social capital can facilitate proactive behaviour (such as migration), its exclusionary nature (particularly for bonding social capital) can increase the vulnerability of some groups (e.g. in-migrants) to environmental change or shocks.

## Notes

1 In the 1970s, individuals from Kadalkarai Uuru reported going to Rameswaram to work on trawlers because the wages were better. However, conflicts with the Sri Lankan coast guards and an increase in larger boats (and the associated increase in fishing incomes) led to a reduction in the migration stream. By the early 1980s, male Thoothoor fishers were already moving farther north up the west coast, to Gujarat and Maharashtra for shark fishing, and stayed away from home for the entire fishing season.
2 All study participants' names are pseudonyms.
3 Worms are available in coastal stretches of marshlands, the mouths of estuaries, and areas close to salt pans and mangroves.

## References


Baas, M. (2017). *What determines the cost of migration? A perspective from Indian agents facilitating migration to Singapore and the Middle East*. Asia Resource Institute Working Paper No. 264. Retrieved from https://ari.nus.edu.sg/wp-content/uploads/2018/10/wps18_264.pdf

Bavinck, M. (2008). Collective strategies and windfall catches: Fisher responses to tsunami relief efforts in South India. *Transforming Cultures*, *3*, 76–92. Retrieved from https://epress.lib.uts.edu.au/journals/index.php/TfC/article/view/923 accessed 5/2/2020.

Bavinck, M. (2014). Investigating poverty through the lens of riches – Immigration and segregation in Indian capture fisheries. *Development Policy Review*, *32*(1), 33–52.

Bhathal, B. (2014). *Government-led development of India's marine fisheries since 1950: Catch and effort trends, and bioeconomic models for exploring alternative policies*. Unpublished doctoral thesis, University of British Columbia, Vancouver.

Bourdieu, P. (1977). *Outline of a theory of practice* (R. Nice, Trans.). Cambridge: Cambridge University Press.

Cleaver, F. (2005). The inequality of social capital and the reproduction of chronic poverty. *World Development*, *33*(6), 893–906.

de Haas, H. (2010). The internal dynamics of migration processes: A theoretical inquiry. *Journal of Ethnic and Migration Studies*, *36*(10), 1587–1617.

Garip, F. (2008). Social capital and migration: How do similar resources lead to divergent outcomes? *Demography*, *45*(3), 591–617.

Goulden, M.C., Adger, W.N., Allison, E.H., and Conway, D. (2013). Limits to resilience from livelihood diversification and social capital in lake social – Ecological systems. *Annals of the Association of American Geographers*, *103*, 906–924.

Granovetter, M.S. (1973). The strength of weak ties. *American Journal of Sociology*, *78*(6), 1360–1380.

Jordan, J.C. (2015). Swimming alone? The role of social capital in enhancing local resilience to climate stress: A case study from Bangladesh. *Climate and Development*, *7*(2), 110–123.

Massey, D., and García-España, F. (1987). The social process of international migration. *Science*, *237*, 733–738.
Molyneux, M. (2002). Gender and the silences of social capital: Lessons from Latin America. *Development and Change*, *33*(2), 167–188.
Palloni, A., Massey, D., Ceballos, M., Espinosa, K., and Spittel, M. (2001). Social capital and international migration: A test using information on family networks. *American Journal of Sociology*, *106*, 1262–1298.
Portes, A. (1998). Social capital: Its origins and applications in modern sociology. *Annual Review of Sociology*, *24*, 1–24.
Rajan, J.B. (2002). *Labour mobility in the small-scale fisheries sector of Kerala*. Discussion Paper No. 44. Thiruvananthapuram: Centre for Development Studies.
Ram, K. (1991). *Mukkuvar women: Gender, hegemony and capitalist transformation in a south Indian fishing community*. London: Zed Books.
Shelton, C. (2017). *The role of culture in adaptive responses to climate and environmental change in a Fijian village*. Unpublished doctoral thesis, University of East Anglia, Norwich.
Silvey, R., and Elmhirst, R. (2003). Engendering social capital: Women workers and rural-urban networks in Indonesia's crisis. *World Development*, *31*(5), 865–879.
Smith, A. (2000). Mobilizing social resources: Race, ethnic, and gender differences in social capital and persisting wage inequalities. *Sociological Quarterly*, *41*(4), 509–537.
Vivekanandan, E. (2011). *Climate change and Indian marine fisheries*. Marine Fisheries Policy Brief No. 3. Central Marine Fisheries Research Institute Special Publication No. 105. Kochi: Central Marine Fisheries Research Institute.
Woolcock, M., and Narayan, D. (2000). Social capital: Implications for development theory, research, and policy. *World Bank Research Observer*, *15*(2), 225–249.

# 10 Mobilizing for and against migration

## Gendered networks, cooperation, and collective action in fishing communities on the west and east coasts of Sri Lanka

*Nirmi Vitarana, Dilanthi Koralagama, Nireka Weeratunge and Ramani Gunatilaka*

### Introduction

Social relations and networks constitute an integral aspect of migration among fishing communities in Sri Lanka. Historically, fishing households have adopted migration as a livelihood strategy to overcome seasonal vulnerability caused by monsoonal weather patterns, which prevent regular fishing activities in their home villages. Fishers' migrations involve collectives comprising fishing households, which mobilize their material resources (economic assets) and relational resources (social networks) to work during a fishing season in host locations. Thus, social relations that are unique to men and women in fishing communities and networks are instrumental in supporting or constraining migration. This chapter analyses the significance of social relations and networks in enabling or disabling seasonal migration in four small-scale fishing communities on the west and east coasts of Sri Lanka. We argue that although social networks provide emotional and crisis support in all four communities, the degree of livelihood support from social networks is higher for both men and women in west coast communities than in east coast communities. Stronger support for livelihoods, along with stronger bridging and linking networks, enable west coast fishing communities to mobilize for migration and engage in collective action to safeguard their right to migrate, despite contestation by host communities on the east coast.

To contextualize the role of social relations and networks in migration, this chapter begins with an overview of the sociocultural composition (ethnicity, religion, and caste) of fishing communities in Sri Lanka, as well as historical changes in migration processes (elaborated in Chapters 4 and 6, this volume). Thereafter, the conceptual approach and methods are outlined, followed by the analysis of the types of social relations, the significance of social networks for livelihood support, and how both the relations and the support are mobilized for internal seasonal migration. The chapter concludes with a summary of the main findings and arguments relating to the role of networks in enabling cooperation and collective action both for and against migration among fishing communities.

## Sociocultural and historical overview

Historically, fishing in Sri Lanka has been associated with particular castes. The most recent study of caste composition in fisheries (Munasinghe, 1985) revealed that almost 80% of fishers belonged to the Karava (Sinhalese)/Karaiyar (Tamil) caste. In terms of ethnicity, 54% were Sinhalese, 36% were Tamils, and 10% were Muslims and Malays. The majority of fishers were Catholics (46%) by religion and ranged across both the Sinhalese and Tamil ethnic groups. While these socio-cultural characteristics might have changed somewhat with the acquisition of multiday fishing boats by entrepreneurs from other social groups, most fishers who fish at sea are from castes customarily associated with fishing. On the west coast of Sri Lanka, ethnic boundaries between the Sinhalese Catholics and Tamil Catholics belonging to the Karava/Karaiyar and Paravar castes (all with southern Indian ancestry) are fuzzy, and many from these social groups have maintained a bilingual hybrid identity into the 21st century.

Fishing communities in Sri Lanka have resorted to internal migration from coast to coast and along the coast, primarily in response to monsoon wind and rain patterns and to fish movements. Historical records of migrant fishers from the Negombo area on the west coast having moved seasonally to Mullaitivu District, Trincomalee District and Batticaloa District on the north and east coasts exist in colonial administration reports dating from 1868.[1] Currently, migrants can recollect accounts of migration patterns given by their parents and/or grandparents. Migrants were of the same castes or similar castes (i.e. Karava/Karaiyar or Paravar), spoke Tamil (the language of the host communities on the east and northeast coasts), and interacted with local people while practising a common livelihood. Some intermarried with local people and established kin relations, while others had a family on each coast. However, migration patterns from the west coast to the east and northeast coasts have been negatively affected by almost three decades of civil war (1983–2009), with both economic and social consequences. Households in both study districts (Puttalam and Trincomalee) suffered not only the violence and trauma of the ethnic conflict but also the 2004 tsunami, both of which caused the loss of family members, livelihood assets, and capabilities, such as fishing skills, and severely hampered fishing activities. The return of large numbers of migrant fishers and boats to the east and northeast coasts since the war has heightened tensions between migrant communities and host communities. Despite similarities in caste and language, identities have increasingly become defined by the geopolitical region to which fishers belong, and divisive notions of 'us' and 'them', and 'north' and 'south', are beginning to impede previous coast-to-coast pathways.

## Conceptual approach

The conceptual approach adopted in this chapter is informed by the interface of the concepts of social relations, social networks and capital, and social well-being. In the study on which this chapter is based, social relations were analysed within an institutional framework, which included four domains: family and kinship, community, market, and the state (Kabeer, 1999). Gender relations fall under these broader social relations, and according to Kabeer, 'like all social relations, [they] are constituted through the rules, norms and practices by which resources are allocated, tasks and responsibilities are assigned, value is given and power is mobilized' (Kabeer, 1999, p. 12). Additionally, social networks, as patterns of linkages and relationships among individuals in households, were explored across the aforementioned four domains: family and kinship, community, market, and the state. These networks are important determinants of the social resources available to households to pursue their livelihood activities. As also discussed in Chapter 9, social networks, as defined here, are close to the concept of social capital, as originally proposed by Bourdieu (1986) (i.e. social resources that enable individuals to navigate their position within a hierarchical social structure and provide potential benefits relative to their inclusion in such networks). Thus, the strength or weakness of social capital entails social inclusion or exclusion respectively. However, the concept of social networks, rather than social capital, was preferred for our analysis, due to the narrower, functionalist, and utilitarian turn taken towards the latter concept in the work of Coleman (1988) and Putnam (2000), as well as a reluctance to reduce an intricate web of social relations to an economist notion of capital.

Among numerous benefits, social networks provide food, shelter, finance, labour, care, moral support, tacit 'cultural knowledge', access to education and training, information and employment, and status and identity, all of which can have variable impacts on the livelihoods and well-being of households. Social networks also exact costs, such as obligations to others, especially in times of celebration or misfortune. While these relationships of reciprocity offer security and help to mitigate risk, they can also maintain the poor in social structures that perpetuate poverty (Wood, 2003) or entail conflicts, rather than trust and cooperation.

The role of social capital or networks, especially that of bridging and linking ties in fisheries governance, has been discussed widely in the global literature (Bodin and Crona, 2009; Grafton, 2005; Holland et al., 2013; Marín et al., 2012; Nenadovic and Epstein, 2016; Sekhar, 2007). Additionally, the relationship between social networks and migration (Kraan, 2009; Overå, 2001) has been addressed in the fisheries literature, but less extensively. Siriwardena-de Zoysa (2018) elaborates on the significance of cooperation that is based on social ties, referred to as *sambandam* in Tamil (*sambandatha* in Sinhalese), in the everyday lifeworlds of coastal groups in Trincomalee

District. While remaining critical of Putnam's approach to social capital (Putnam 2000), we adopt his typology of bonding, bridging, and linking ties, as defined by Nenadovic and Epstein (2016), in our analysis of social networks that support or constrain livelihoods and migration: relationships among individuals with similar demographic characteristics, such as family members, friends, and neighbours, are termed *bonding*; relationships among individuals that differ in these characteristics but live in proximity to one another are termed *bridging*; and relationships among individuals that differ in their positions of authority, such as between fishers and state officials, are termed *linking*.

Further, the relationship between social capital and well-being has been discussed in the fisheries literature (Fowler and Etchegary, 2008; Isham, 2000). The relational dimension is fundamental to the concept of social well-being, alongside the material and subjective dimensions (McGregor, 2008; White, 2008). White (2008) explains social well-being as *having* a good life, which is defined by material attributes and standard of living; *living* a good life, in which relationships and capabilities play an important role; and *locating* one's life with reference to subjective notions of happiness. Thus, in the approach presented in this chapter, migration is regarded as enabled or disabled by social relations and networks constituted by the bonding, bridging, and linking ties of members of fishing households in the pursuit of their well-being.

## Methods and study locations

Data collection, based on mixed methods, was conducted in four fishing villages in 2016 and 2017:[2] Kadalpalli and Cattiyur[3] in Puttalam District, on the west coast, and Selippur and Uppukadal in Trincomalee District, on the east coast of Sri Lanka. Details of the study participants are listed in Table 1.1 in Chapter 1, and details of the methods are given in Chapter 6. This chapter is based on qualitative social network mapping undertaken to identify types of social relations and networks, corroborated by quantitative data from a household survey conducted to analyse the importance of networks for livelihood support. Additionally, qualitative data from focus group discussions (FGDs) and livelihood profiles and life histories of women and men are used to explore how networks are mobilized to enable or disable migration.

The two sets of villages comprised sending (migrant) communities and receiving (host) communities respectively. Kadalpalli is a predominantly Sinhalese Catholic village, in which most households engage in small-scale marine fishing, using small motorized fibre-reinforced plastic boats.[4] Cattiyur is predominantly a Tamil Hindu village, in which fishers use motorized boats and traditional vessels (rafts), and a relatively large labour force is engaged in beach seine fishing. Selippur is a predominantly Muslim village, in which marine and lagoon fishing with motorized boats is combined with some farming. Uppukadal is a predominantly Tamil Hindu village, with households engaged in marine and lagoon fishing that uses motorized boats and traditional vessels (canoes) and in non-fisheries work.

## Types of social relations and networks

Qualitative social network mapping in the four study villages revealed that four categories of social relations and two types of networks based on those relations were instrumental in supporting the livelihoods of women and men:

1. relations with family, friends and neighbours
2. relations in community organizations and institutions (e.g. fisheries societies, women's societies, village church, temple, and/or mosque)
3. labour relations (e.g. relations between boat owners and crew, between fish processors and employees, and between coworkers)
4. market relations (relations with traders, namely suppliers of inputs and buyers of products, and relations with banks).

Social networks based on the first two types of relations can be conceptualized as being constituted of 'bonding ties', while those based on the second two types can be categorized as being constituted of 'bridging ties'. However, according to each village context, relations in community groups and labour relations can also cut across the 'bonding' and 'bridging' forms, thus reinforcing Putnam's observation that the two types of ties are not mutually exclusive (Putnam, 2000).

### *Relations with family members, friends, and neighbours*

The strongest linkages for both men and women from all four study villages were relations with family members, friends, and neighbours. Both women and men depended on the bonding networks of family members, friends, and neighbours in their own villages to support their livelihood activities. As the kinship pattern in three of the villages (Cattiyur, Kadalpalli, and Selippur) is predominantly matrilocal, these relations were primary for women. Apart from family members, men identified strong relationships with friends in their village. Men who had moved into the village from outside (mainly through marriage) also had strong relations with friends and family members who lived outside the village and sometimes in other countries. In contrast to men, only a minority of women had linkages with family members or friends from outside their village or district.

### *Relations in community organizations and institutions*

Relations in community organizations and institutions were among mainly homogenous groups bound by a common location, identity, and reciprocal ties in the four study villages. These groups included cooperative fisheries societies, women's groups, savings and credit societies (some initiated by government, non-governmental organizations, or private leasing and/or credit companies) and local places of worship, such as churches and temples. Among these groups, networks associated with fisheries cooperative societies were the most important, especially for male fishers, who assessed the links as medium to strong in the social network mapping done in the west coast villages. The societies provided access to loans for fishing activities and mobilized members for rescue assistance when fishers were lost at sea. The high degree of cooperation and reciprocity and the ability for collective action in times of distress were important values in all study villages. Women were generally excluded from membership or participation in fisheries societies but were members of other community-based organizations that are specifically intended to serve them, such as savings and credit groups. Women who belonged to such groups supported their husband's livelihood activities by obtaining loans for new fishing gear; for the repair of fishing boats, engines, nets, and longlines; and for financing their own fish processing or trading activities.

For both women and men, social networks associated with religious institutions, especially the village church, temple, or mosque, were strong. Regular religious rituals and annual festivals are important for maintaining a community's social cohesion and identity as a fishing community and form the underlying basis of social relations and networks. Several annual rituals were performed to ensure the success of fishing activities, especially at the beginning of the fishing season. While both women and men participated in the religious life of their community, men's links with religious institutions were generally stronger in the Hindu and Muslim communities, whereas women and men had equally strong links with the church in the Catholic community in Kadalpalli. In the same village, women were also leaders of *novena*[5] groups and initiated some religious activities, mainly for congregations of women. In the west coast villages, the church and the largest Hindu temple were directly involved in livelihoods in that they owned and/or operated the main local marketplace for fish trading. The local religious institutions thus supported networks where bonding ties were nurtured and bridging ties were enhanced through linkages facilitated by local religious leaders connected to translocal networks.

### *Labour relations*

In the four study villages, labour relations were determined by the types of fishing vessel and fishing gear used. Those who used rafts or canoes either fished alone or relied on a family member, friend, or neighbour to be a helper. Those who operated motorized boats fished with one or two crew members, who could be kin, a neighbour, or an outsider to the village. Boat owners who did not fish employed two or three fishers as crew. In all three contexts, the helpers and/or crew were paid a share of the catch as wages, according to established norms in the community. Regardless of whether kin or non-kin were employed, the structural pattern in the relations was that of an employer and employee, with prescribed payments. Thus, depending on the specific relationship between the boat owner and the helper or crew member, the associated social networks can be considered as having had a range of bonding and bridging ties, with mutual dependency in pursuing fishing as a livelihood and maximizing earnings. While relations between boat owners and crew were cordial, boat owners sometimes displayed a paternalistic attitude towards their crew.

Unlike the owners of motorized boats, beach seine owners require a larger number of labourers, ranging from 20 to 40. These workers often come from their home villages or from Batticaloa District on the east coast, thus constituting both bonding networks and bridging networks. Beach seine owners, while maintaining a social distance between them and their labourers, said that they treated the labourers well and made efforts to keep them happy because reliable labour was critical to the success of the beach seine operations. They often expressed paternalistic attitudes towards their workers. Some workers in Cattiyur depicted beach seine work and their relations with beach seine owners as 'slavery', especially those who had left the beach seine fisheries when they or their family members had migrated overseas for work.

Women were also employed in beach seine *wadi* (fishing camps) for fish sorting and processing. Relations between female workers and the beach seine owners were considered cordial, based on respect, especially on the part of the workers, who often negotiated through the wife of the beach seine owner. Female workers who worked together in fish sorting and/or processing formed small cliques, relying on one another for childcare, food sharing, and borrowing money. While these peer groups were based mainly on reciprocity and solidarity among networks of kin, friends, and neighbours, they sometimes also entailed ties with women from outside the village. Most labour relations among both men and women were associated with bonding networks. However, men were more likely to be involved in bridging networks with their employers and coworkers, some of whom were from outside the village, and thereby expanded their opportunities for livelihoods and income.

### *Market relations*

Market relations in the four study villages included bridging linkages inside and outside the village with fresh-fish traders and dried-fish traders, as well as suppliers of inputs, such as nets, engines, boats, and salt. These relations cut across different castes and ethno-religious groups (Sinhalese, Tamil, Muslim, Christian, Buddhist, and Hindu). Market links were more complex in the west coast villages because of value chains that were longer than those in the east coast villages. Women's roles included bringing the fresh-fish catches landed by their husbands to the village market to auction in west coast villages and sometimes to trade in neighbouring villages. Local customers and small traders from the village and small-scale, medium-scale, and large-scale traders of different ethnic groups and castes from outside the village arrived at the market to buy fish. Medium-scale fresh-fish traders (*ice mudalali*), usually of the Karava/Karaiyar and Paravar fisher castes from the village, purchased the fish that remained unsold by 10.00 a.m. Relations between fishers and traders were generally formal, with some ambivalence about fish prices and debts owed by fishers to traders. However, fishers considered certain traders as having their trust and therefore perceived a form of kinship with them. Apart from some fishers who sold their catches to traders who were their kin, most often the relationship between fisher and trader was one of a structurally vertical bridging tie, although it was sometimes imbued with an appearance of a horizontal bonding tie. Fish traders perceived relations between fish traders and fishers as mutually beneficial and cordial. This was also the case for female micro- and small-scale fish traders.

Most fishers in the east coast villages were 'tied' to local traders who provided loans to purchase boats and fishing gear, because the loans were raised against their future catches. Fresh-fish catches on the east coast were sold mainly directly on the beach by fishers to medium-scale traders in the village and to medium-scale and large-scale external traders, who came from different parts of the country. Few women in east coast villages were involved in fish trading in the village, while some women processed dried fish and sold it to traders from large towns.

### *Relations with the state and the Church*

Apart from the four types of more-widespread social relations discussed thus far, our social network mapping revealed a small minority of households, especially those of community leaders, as having linking ties to officials and personnel in institutions beyond the village, such as fisheries authorities, divisional and district government officials, politicians, and church authorities. These ties were stronger in the west coast villages, especially Kadalpalli, compared with in the east coast villages.

## The significance of social networks in providing support for livelihoods and migration

Both women and men in all four study villages indicated that they could count on relatives, friends, and neighbours for emotional and crisis support, as shown by the quantitative data in Figure 10.1. However, there were noteworthy differences in the importance of social networks that supported livelihoods between east coast villages and west coast villages, and between women and men. While most women and men in the west coast villages indicated that they could turn to relatives and friends in their village for

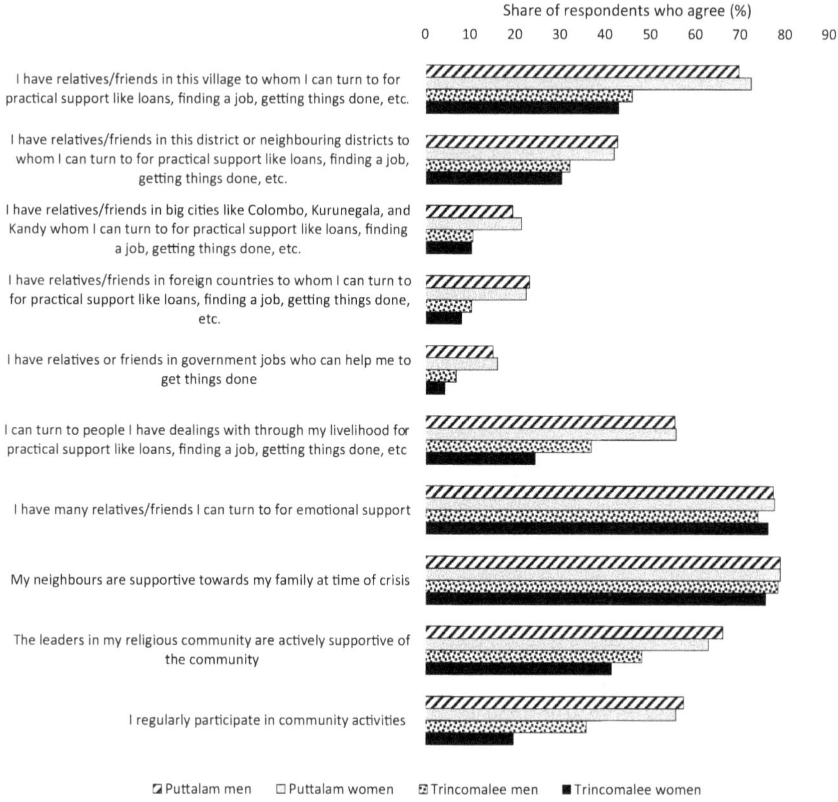


*Figure 10.1* Livelihood support from social networks for women and men in all households in study villages, Puttalam District and Trincomalee District, Sri Lanka, 2017

Source: ICES Survey on Migration and Livelihoods in Fishing Communities 2017

practical support in their livelihoods, only a minority of women and men in the east coast villages were able to do the same. This was also the case for support, which enabled or disabled migration processes. According to data from the household survey in 2017, significantly more men and women from migrating fishing households in Kadalpalli could count on livelihood support from both bonding and bridging networks than could those from non-migrating fishing households, whereas in Cattiyur and Uppukadal the difference between the two types of households was not large. In Selippur, the number of households that migrated for fishing was too small for comparison.

Bonding ties with family, neighbours, and friends were important considerations when household members jointly discussed their migration options and made decisions: 'Everybody in this neighbourhood migrates to one particular place. They collectively decide on it. All are our relatives, actually brothers and sisters. Sharing is not uncommon. We share our meals too'[6] (Lalani, wife of a migrant fisher, Kadalpalli, Puttalam).[7] Crew who accompany migrant boat owners are often relatives or neighbours, and some migrant beach seine owners employ mainly people from their own village:

> Those who migrate from this village stay together in a *wadi*. When we migrate, we do so in batches, so that several lorries leave the village together. Our people have already started to get organized to migrate now.
>
> (Participant in Men's FGD, Kadalpalli, Puttalam)

> I take my workers to Jaffna from Cattiyur. Some people hire workers from Elephant Pass and Batticaloa but I don't.
>
> (Sivarasa, male beach seine owner, Cattiyur, Puttalam)

Women rely on their kin to raise money for their husbands' migration and to support their own livelihoods in the village while their husbands are away:

> We must pay around 100,000 rupees [USD655] in debt every month now. That's why my husband went to Nayaru with the fishers. . . . Even now, my nieces gave us their jewellery to be pawned so that he had enough money to go to Mullaitivu. Next month is the feast at Kotapitiya Church, so we have to recover the jewellery and settle our debts by then. My son and two brothers help to deliver the fish [sent by her husband] to Negombo or Colombo by lorry.
>
> (Darshini, wife of a fish trader and shrimp farmer, Kadalpalli, Puttalam)

The responses to the statement 'I can turn to people I have dealings with through my livelihood for practical support like loans, finding a job, getting things done' indicate that there was considerably more support from work and/or business ties in the west coast villages than in the east coast villages. Whereas the difference between women and men in the west coast villages was negligible, in east coast villages, a larger proportion of men could rely on support than could women. The differences can be explained by the higher numbers of women engaged in fish sorting and fish processing in the west coast villages.

For men from the west coast villages, bridging ties in the form of linkages with traders in host villages on the east coast are critical for facilitating migration. Traders are instrumental as sponsors in advancing the required sums to purchase fishing gear and other supplies to support migration and in providing protection against potential tensions with host communities. The ethnicity of the host traders (Tamil or Muslim) may differ from that of the migrants (Sinhalese or Tamil); the traders nevertheless protect migrant fishers during conflicts that might arise in the places to which they migrate:

> In Uppukadal, a Tamil trader from that village is involved [as a sponsor]. He comes before the season starts and advances us money to migrate. This trader advances fuel expenses and credit for migration, and also provides toilet facilities [in the migrant camp].
>
> (Participant in men's FGD, Kadalpalli, Puttalam)

> The *mudalali* [trader] for whom we work intervenes and resolve issues if they come up. The *mudalali* is a Muslim from that area. If needed, it is he who resolves issues by taking them up, even up to the level of the police.
>
> (Participant in men's FGD, Kadalpalli, Puttalam)

However, there was considerable ambivalence in the fishers' relations with traders, due to dependency and indebtedness, and fishers relied on peer support if debts owed to traders posed a serious threat to the continuity of their livelihoods:

> Before fishers migrate from the village during the monsoon, they have a lot of expenses to bear in preparation. Although the *mudalali* who sponsor the fishers advance money, if fishers are not able to catch enough fish to pay them back, there are some *mudalali* who take possession of the boat engine as collateral before they [the fishers] return back to the village. So, fishers tend to live in a cycle of loans and debts. There are both considerate *mudalali* and also non-flexible, money-driven *mudalali*. When a fisher gets into a cycle of never-ending debt, usually friends come together and help out. That type of camaraderie is there among us.
>
> (Nicholas, male commission agent,[8] Kadalpalli, Puttalam)

In addition, while most men and women participated in community activities in the west coast villages, only a minority did so in the east coast villages. Moreover, most participants from the west coast villages acknowledged support from religious leaders to the community, while a minority did so in the east coast villages. While the gender difference with respect to participation in community activities and receiving support from religious leaders was not large in the west coast villages, more men than women in east coast villages reported that they participated in community activities and received support from religious leaders. Thus, livelihood support from both bonding and bridging networks emerged as stronger in the west coast villages than in the east coast villages. Moreover, while the difference between men and women in access to both bonding and bridging networks in the west coast villages was not large, there was a considerable gender gap in the east coast villages.

In common with all other households in the study villages, fishing households on the west coast indicated that they received more livelihood support from social networks in general than did fishing households in the east coast villages (Figure 10.2). In this regard, all support from family, friends, neighbours, and relatives was categorized as 'kin', all support from work-related relations (encompassing both labour and market relations) was categorized as 'work', and all support from religious leaders in the community was categorized as 'religious'.

Considerably more livelihood support from bonding networks (kin) and bridging networks (work-related relations and religious leaders) was provided to women and men in fishing households in Kadalpalli than in all other villages. Most women and men from fishing households in Cattiyur

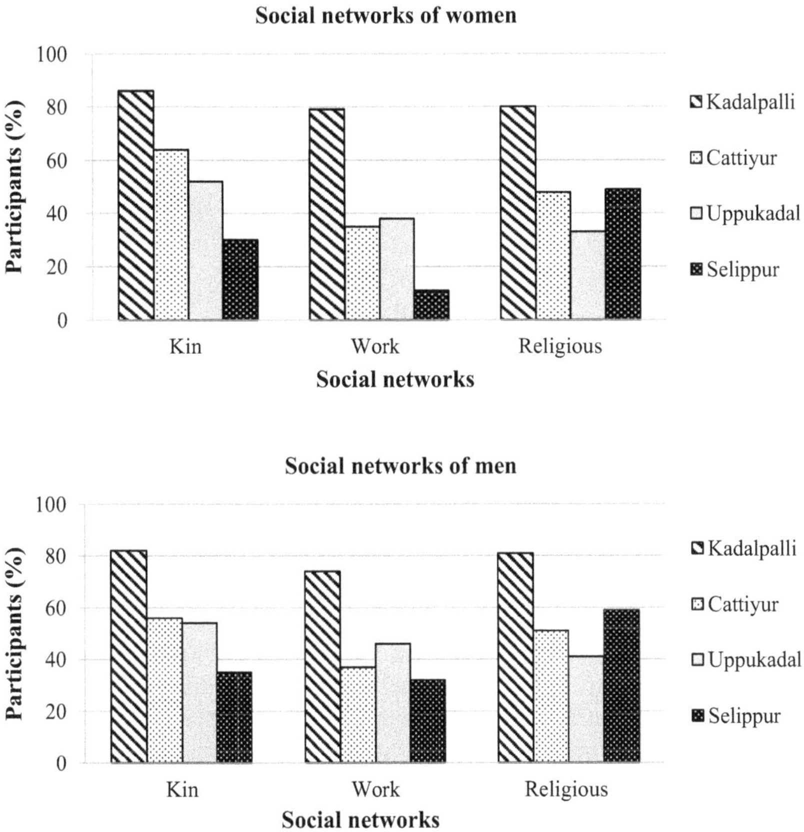


*Figure 10.2* Livelihood support from social networks for women and men in fishing households in study villages, Puttalam District and Trincomalee District, Sri Lanka, 2017

Source: ICES Survey on Migration and Livelihoods in Fishing Communities 2017

could count on kin and religious leaders for livelihood support, but only a minority could count on work-related contacts. Relatively more male fishers in Uppukadal could count on work-related contacts than in Cattiyur and Selippur, while relatively more male fishers in Selippur could count on religious leaders than in Cattiyur and Uppukadal. In the west coast villages, there were no differences between women and men in fishing households in terms of accessing livelihood support from networks, while there was a considerable difference in the east coast villages, where men in fishing households received more support than did women in the fishing households. The weaker networks appear to indicate the social exclusion of east coast fishing communities in general and of women in particular.

The difference between Kadalpalli and the other three villages is probably based on cultural and economic factors. Kadalpalli has a strong identity as a Roman Catholic village, where the Church and religious networks play a central role in the community. The significance of the Church has also been noted in the social networks of fishing households in Kanyakumari, India (see Chapters 7 and 9, this volume). Moreover, the majority of fishing households in Kadalpalli owned their own motorized boats, while only a minority did in the other three villages. Ethnicity, religion, and class appear to have contributed to stronger networks for livelihood support and social inclusion among households in Kadalpalli compared with the other three study villages.

## Mobilizing social networks for collective action in seasonal migration

While a range of community-based associations existed in all four villages, fisheries societies emerged as most important for engaging in collective action, albeit sporadically. Fishers in the two west coast villages mobilized for collective action through their membership of fisheries societies to negotiate with fisheries and local government authorities, and the armed forces at their migration destination, as well as when faced with counter-challenges from host fisheries societies:

> We get a letter from our [fisheries] association and another letter from the Grama Niladhari [village officer] prior to migration, then it is easier to solve any conflict or administrative problems that might come up. Last time, also one boat was burnt by local people at the site of our *wadi*. It was left behind by the owner. We all got together and collectively bought a boat for him.
>
> (Participant in men's FGD, Kadalpalli, Puttalam)

Fisheries societies in the east coast villages are weaker because of limited membership and resources. Their linkages with and support from the Sri Lankan government's Department of Fisheries and Aquatic Resources (in the Ministry of Fisheries and Aquatic Resources Development) and the armed forces appear to be weaker too:

> The navy and army are all on the migrants' side. The navy took us to the police, accusing us of using dynamite to catch fish that we had caught with hook and line. The police told them that this was not dynamited fish.
>
> (Participant in men's FGD, Uppukadal, Trincomalee)

However, there were occasions when fisheries societies in Trincomalee District managed to unite and mobilize collective action on migration issues with the help of local politicians, especially to demand a stop to seasonal migrant fishers from accessing fish resources from their coast:

> We do not have enough space and fish in the sea due to migrant fishers. We do not have any jobs other than fishing. We participated in a demonstration against them but we could not stop them. They have political power and patronage.
>
> (Ruthrakumar, male fisher, Uppukadal, Trincomalee)

The access to politicians was confirmed by a representative of a fisheries society in Kadalpalli, who referred to the quotas imposed on boats and the subsequent response by migrant fishers:

> When 37 fisheries associations in Trincomalee got together to stop us in 2015, the GA[9] gave approval for 200 boats. The local fisheries organizations even requested the transfer of the DFA [District Fisheries Assistant Director, the chief fisheries officer for the district]. Having got the approval [to migrate] at ministry and department level, we went to the church and explained the situation. The bishop advised the local people [in the host villages] not to obstruct our arrival. The Catholic priest in Kadalpalli helped us a lot. . . . When we have a problem with local people [in the host villages], we go to the bishop. The Bishop of Chilaw informs the Bishop of Trincomalee. He provides us with the necessary documents and guidance.
>
> (Dinesh, male migrant fisher and fisheries society representative, Kadalpalli, Puttalam)

Thus, weaker bridging ties and linking ties among east coast fishers reveal higher levels of social exclusion in an institutional context. West coast fishers have been able to use stronger linking social networks with fisheries and church authorities to mediate and counter pressure from host communities. The migrant fishers' practice of migrating and living at the destination sites as large collectives enables individual fishers to come together as a 'set of interacting actors' (Bodin and Crona, 2009) who mobilize if they face external threats. While bridging networks associated with formal fisher associations and religious institutions provide a degree of protection and solidarity, both men and women who face adversity rely first and foremost on their bonding networks of family, relatives, and neighbours who migrate with them.

## Conclusions

Social relations and networks are important for enabling and disabling seasonal coast-to-coast fisheries migration in Sri Lanka. While social networks provide critical emotional and crisis support in all four study villages, livelihood support provided by social networks is stronger in west coast study villages, especially in the case of Kadalpalli. Furthermore, while there are no significant gender differences in support from both bonding and bridging networks in the west coast villages, bridging networks provide more support to men than to women in the east coast villages. Both ethno-religious factors and class factors seem to contribute to the strength of social networks in providing livelihood support in Kadalpalli.

Relatives, neighbours, and friends play a significant role in turning individual fishers and households into a collective force during the migration process in west coast communities. Female migrants also depend on these bonding networks for childcare and informal loans at the destination sites, where they support their husbands in fishing and engage in fish-processing activities. Men depend on cooperative fisheries societies to negotiate access to migration through the formal issuance of relevant letters and to mitigate potential threats from host communities. They also depend on fish traders, mostly from host communities, to facilitate financial assistance and sponsorship for migration, as well as market linkages. Traders play a critical role in negotiating with local communities and authorities in the migrants' destination locations. In addition to these bonding and bridging ties of cooperation, which are important in the everyday lifeworlds of fisher groups, as elaborated by Siriwardena-de Zoysa (2018), fishers also sporadically organize collective action through their cooperative fisheries societies, such as demonstrations and protests for and against migration, and mobilizing the support of linking networks. Fisheries society leaders on the west coast rely on ties with regulatory authorities, such as the Department of Fisheries and Aquatic Resources, politicians, and the Church to negotiate on their behalf if they are confronted with opposition from east coast communities and authorities. Similarly, fisheries society leaders on the east coast seek to gather support from local authorities and politicians to curb migration from the west coast, albeit with limited success, due to their relative social exclusion, especially by the state and powerful religious bodies in an institutional context.

Thus, compared with households on the east coast, the stronger bonding, bridging, and linking networks of west coast fishing households enable seasonal coast-to-coast migration. While east coast fisheries societies have been able to limit numbers of migrants through collective action to some extent, migrant fishers have largely been able to withstand pressure from local communities, due to their stronger social networks.

## Notes

1 R.L. Stirrat, 1972, notes on Ceylon Administration Reports, 1867 – 1887, held in personal archive material.
2 These constitute four Grama Niladhari (GN) divisions, the smallest administrative units in Sri Lanka.
3 The names of all study villages are pseudonyms, in accordance with ethnographic convention.
4 In Sri Lankan fisheries reports and statistics, the abbreviation OFRP (outboard-engine fibreglass reinforced plastic) boat is commonly used to refer to small motorized fibre-reinforced plastic boats, and the abbreviation NTRB (non-motorized traditional boat) is used to refer to traditional canoes and rafts.
5 A *novena* is a form of Catholic prayer that continues for a series of nine days to obtain special blessings or favours or to make a special petition by honouring statues of Jesus Christ, the Virgin Mary, or saints.
6 The quotes from study participants in Kadalpalli village were translated from Sinhala into English, while all other quotes from study participants were translated from Tamil into English. The translations were done by the researchers who conducted the interviews, among whom were the authors of this chapter.
7 The names of all study participants are pseudonyms, in accordance with ethnographic convention.
8 A commission agent purchases fish from village fishers for external traders.
9 The research participant referred to the district secretary by the old administrative term government agent.

## References

Bodin, Ö., and Crona, B.I. (2009). The role of social networks in natural resource governance: What relational patterns make a difference? *Global Environmental Change*, *19*(3), 366–374.

Bourdieu, P. (1986). The forms of capital. In J.G. Richardson (Ed.), *Handbook of theory and research for the sociology of education* (pp. 241–258). New York: Greenwood.

Coleman, J.S. (1988). Social capital in the creation of human capital. *American Journal of Sociology*, *94*, 95–120.

Fowler, K., and Etchegary, H. (2008). Economic crisis and social capital: The story of two rural fishing communities. *Journal of Occupational and Organizational Psychology*, *81*(2), 319–341.

Grafton, R.Q. (2005). Social capital and fisheries governance. *Ocean and Coastal Management*, *48*, 753–766.

Holland, D.S., Kitts, A.W., Da Silva, P.P., and Wiersma, J. (2013). Social capital and the success of harvest cooperatives in the New England groundfish fishery. *Marine Resources Economics*, *28*(2), 133–153.

Isham, J. (2000). *Can investments in social capital increase well-being in fishing communities? A theoretical perspective for assessing the policy options*. Unpublished manuscript, Middlebury College, Vermont, USA.

Kabeer, N. (1999). From feminist insights to an analytical framework: An institutional perspective on gender inequality. In N. Kabeer and R. Subramanian (Eds.), *Institutions, relations and outcomes* (pp. 3–48). New Delhi: Kali.
Kraan, M.L. (2009). *Creating space for fishermen's livelihoods: Anlo-Ewe beach seine fishermen's negotiations for livelihood space within multiple governance structures in Ghana*. African Studies Collection 19. Leiden: African Studies Centre.
Marín, A., Gelcich, S., Castilla, J.C., and Berkes, F. (2012). Exploring social capital in Chile's coastal benthic co-management system using a network approach. *Ecology and Society*, *17*(1), 13. Retrieved from http://dx.doi.org/10.5751/ES-04562-170113
McGregor, A. (2008). *Well-being, poverty and conflict*. ESRC Research Group on Well-being in Developing Countries, Briefing Paper 1/08. Retrieved from www.bath.ac.uk/soc-pol/welldev/research/bp/bp1-08.pdf
Munasinghe, H. (1985). Socio-economic conditions of small-scale fishermen in Sri Lanka. In T. Panayotou (Ed.), *Small-scale fisheries in Asia: Socioeconomic analysis and policy* (pp. 73–83). Ottawa: International Development Research Centre.
Nenadovic, M., and Epstein, G. (2016). The relationship of social capital and fishers' participation in multi-level governance arrangements. *Environmental Science and Policy*, *61*, 77–86.
Overå, R. (2001). *Institutions, mobility and resilience in Fante migratory fisheries in West Africa*. CMI Working Paper WP 2001:2. Bergen: Chr. Michelsen Institute.
Putnam, R.D. (2000). *Bowling alone: The collapse and revival of American community*. New York: Simon & Schuster.
Sekhar, N.U. (2007). Social capital and fisheries management: The case of Chilika Lake in India. *Environmental Management*, *39*(4), 497–505.
Siriwardena-de Zoysa, R. (2018). *Fishing, mobility and settlerhood: Coastal socialities in postwar Sri Lanka*. Cham: Springer International.
White, S.C. (2008). *But what is wellbeing? A framework for analysis in social and development policy and practice*. Unpublished manuscript, University of Bradford. Retrieved from http://people.bath.ac.uk/ecsscw/But_what_is_Wellbeing.pdf
Wood, G.D. (2003). Staying secure, staying poor: The Faustian bargain. *World Development*, *31*(3), 455–471.

# 11 Gender and power struggle in community fisheries in Cambodia

## Creating space for women's leadership

*Kyoko Kusakabe, Prak Sereyvath, Lam Doeurn and Yem Sivon*

### Introduction

Since 2002, Cambodia has been encouraging and developing community fisheries (CFs)[1] to provide rights to communities in the management and use of fisheries resources. Kurien (2018) hails this as one of the most progressive resource management paradigms in the world. In 2012, the prime minister of Cambodia abolished private fishing lots and allowed small-scale fishers to fish without any restrictions.

Despite their rights to the management and use of fisheries resources, small-scale fishers have experienced threats from outside investors and are currently experiencing changes in their livelihood options. This chapter presents two cases of community fisheries in the coastal region of Cambodia and analyses how fishing communities in the region have fought for their rights to fishing resources and how their fight has changed over time and analyses the changes in women's involvement in the struggle and the impact of those changes. We explore the changing focus of community fisheries and how women have created space in a male-dominated institution.

### Community fisheries in Cambodia

There are currently 516 CFs registered in Cambodia at the time of this study: 477 inland CFs and 39 marine CFs. Each CF has five to 11 committee members, as an odd number is specified in the 2005 subdecree on the management of community fisheries (Kingdom of Cambodia, 2005). Furthermore, each CF is registered specifically for the area it manages. According to the subdecree issued in 2005, CFs can register for rights and responsibilities over fishing areas, mangrove forests, and inundated forests, where they are to manage fishing resources. In Cambodia, the Fisheries Administration (in the Ministry of Agriculture, Forestry and Fisheries) organizes meetings to disseminate information about the roles and responsibilities of CFs.

Additionally, CFs have horizontal networks that enable them to learn from each other.

Fishing lots were introduced in 1900 during the French colonization of the country. After Cambodia gained independence in 1953, the fishing lots continued to exist. During the Democratic Kampuchea period from 1975 to 1979, the fishing lot system was abolished. After that period, many people moved to find a new livelihood by fishing in Tonle Sap, since the resources were abundant in the lake and it was easy for people to move there. During the People's Republic of Kampuchea in the 1980s, fishing lots were revived and given to groups of fishers to manage. In 1987, private fishing lots were introduced, and private owners paid the government for the rights to them (Hutson, 2006). In the 2000s, violent conflicts started between the owners of fishing lots and small-scale fishers over the fish resources in Tonle Sap. The small-scale fishers' fishing grounds were limited, and the owners of the fishing lots did not allow them to enter their lots. During the conflicts, fisheries officials sided with the fishing lot owners and confiscated equipment from the small-scale fishers. As the conflicts worsened, Prime Minister Hun Sen said in his speech on 24 October 2000 in Siem Reap, 'Government fisheries officials are the bloodsuckers and the guard dogs of fishing lot owners'.[2] In December 2000, the government stopped providing rights to the owners of private fishing lots, and in 2012, the fishing lot system was abolished. Some lots were handed over to CFs for management, and some remained as conservation lots.

The subdecree on the management of community fisheries issued in 2005 was intended to strengthen local capacity in managing fishing resources. The main role of CFs, as instituted by the subdecree, was to patrol the lots in order to protect and manage fishing resources. As fishing resources have declined (as described in Chapter 3, this volume), violent competition over the resources has increased. In addition to the competition over fishing resources, there are also competing claims for land around the water bodies for both tourism and manufacturing industries.

Although the subdecree on community fisheries focuses on fisheries resources, the use of the area in which the resources are located is not regulated for other purposes, such as tourism. Although the subdecree states that CFs have rights, in practice they do not have the power to exercise those rights over their area, which means they cannot control how outsiders use it. CFs are not able to arrest illegal fishers, nor can they automatically claim any control over the use of their area. Rather, they need to go through lengthy court cases to claim their rights over the area. Additionally, the management capacity in most CFs is limited, and given the lack of power, it is difficult to motivate CF members to participate in community-level projects to claim their rights or to fight court cases. Apart from protecting their designated area, CFs are not entitled to engage in any activities other than fishing. According to Chap et al. (2016), the law does not grant the right to CFs to operate commercial activities. As a consequence, the CFs are not assured of the right to claim the use of the area for income-generating activities such as tourism and other types of development. Moreover, the subdecree does not guarantee exclusionary rights to CF, and therefore, even when they make an effort to conserve resources, outsiders can benefit from their efforts.

This chapter explores how CFs have struggled for their rights to resources and how female and male leaders have emerged in the process. First, the analysis framework is introduced. Thereafter, two cases of CFs and their struggles to protect resources are presented. The discussion focuses on a comparison between the two cases and is followed by our conclusions.

## Women in fishing organizations

The establishment of CFs has allowed fishers to take part in the management of fishing resources and directly deal with illegal fishers. It has also increased collective action in some places, especially with the help of civil society (WorldFish 2013). Alonso-Poblacion and Siar (2018) point out that although studies have been conducted on women's participation in fisheries organizations, there has been little research on how women's participation is related to dwindling resources and whether such resources can be an enabler for their participation. Alonso-Poblacion and Siar studied how changes in external environments could create opportunities for women's participation and leadership in fishers' organizations and found that the changes relating to fishing could act as catalysers in women's engagement in collective action, including with respect to dwindling resources and the allocation of fishing rights (Alonso-Poblacion and Siar, 2018). They argue that the crisis faced by fisheries worldwide, which has been triggered by dwindling resources and threats to fishing rights, has forced women to respond collectively to defend their livelihoods. Given the recent decline in fish resources in Cambodia, it is critical to understand its impact on women's participation in CFs.

Women are underrepresented in CFs because fishing is done mainly by men, thus putting women in a secondary position. Women's low levels of participation in fishing organizations means they have comparatively less access to resources and capital, which in turn leads to further neglect and lack of recognition of women in fishing organizations (Alonso-Poblacion and Siar, 2018). Although CFs are supposed to be participatory bodies, they do not necessarily work to change the social order that excludes women and keeps them in a secondary position in the fisheries. Mamun (2015) studied leadership in the co-management of fisheries by the government and communities and identified that the co-management had helped to empower people who were already empowered, and it thus replicated the existing social order. In a review of the literature, Meola (2013) identifies barriers to women's leadership in natural resources management, which include patriarchal norms and religious beliefs that cast doubt on women's ability to lead, marital status, gendered divisions of labour, lack of economic resources, lack of education, lack of confidence, and lack of awareness of opportunities. From her study conducted in Brazil, she found that women's low representation in the management of natural resources was due not only to the lack of opportunity but also to more complex issues, which included some of the aforementioned issues (Meola, 2013). Uchida et al. (2019) note that self-esteem is essential for individuals to take up challenges and that those with high self-esteem tend to participate in collective actions regardless of their class, age, or education levels. Rowlands (1995) emphasizes the importance of self-confidence and having a sense of agency in participating in collective action. However, the just-discussed recommendations do not apply to all people in all contexts (i.e. they are not fixed formulae).

Through the process to overcome the challenges that women face in establishing their leadership, they create their own space for leadership. Sheridan et al. (2011) discuss how businesswomen operate in the 'space of betweenness' – between the private and public spheres and between different roles as managers and mothers – and how this has made their roles invisible. However, such betweenness can also be seen as a strength when women create spaces for themselves. Border studies have shown how geographical margins create a space for the marginalized to claim certain spaces of their own (Donnan and Wilson, 1999; Sturgeon, 2005). Because the margins are far away from the existing power centre, they are able to keep the power at a distance, which Scott (2009, p. 24) refers to as the 'flight zone' that keeps the state at arm's length (p. 62). The marginal space that women are often allotted can be a strength that women can lay claim to and exert leadership in (Cornwall, 2003). Space at the margin can create a space for women to challenge male-dominated organizations. However, at the same time, whether or not such opportunities can lead to even greater power and sustained leadership for women depends on the organizational and socioeconomic environment and on the governance structures (Meinzen-Dick et al., 2006). Women can quickly lose their claim when the marginal space becomes absorbed into mainstream governance. As Khan et al. (2015) and Doss and Meinzen-Dick (2015) note, how women gain power is a circular process that is facilitated or hampered by factors such as the external environment, the governance system, and interactions with others. Higgitt (2011) notes that even when women gain self-confidence and the ability to lead, their leadership will not be realized while their culture context remains unchanged.

This chapter explores how the shift in contested areas led to the rise and fall of women's leadership, through case studies of two community fisheries in two coastal villages. The study was conducted in 2017 and 2018 in K'dam Community Fishery in Koh Kong Province and in Konkang Community Fishery in Kampot Province.[3] Information was gathered through key informant interviews with heads of the two CFs; focus group discussions with women, men, and youths; and in-depth interviews with 13 fishers in Koh Kong Province (nine women) and 11 fishers in Kampot Province (eight women).

## The case of K'dam Community Fishery

Infrastructure in Koh Kong Province was not developed until the early 2000s, when the population density was low. The people were engaged in upland shifting cultivation, because it was possible for villagers to open up as many fields as they needed. They grew upland rice, corn, and pumpkins and were relatively self-sufficient. Each household cultivated approximately 1–3 ha. At that time, fishing was not an important livelihood, since the households were able to source enough food from farming. However, in 2002, economic land concessions were introduced in the area where the villagers were planting upland rice. Upland farms were confiscated by the government for a public good and converted into economic land concessions to be further developed into an industrial area. Most farmers were vulnerable to land confiscation for economic concessions because they did not hold any land titles. However, although some farmers were able to claim ownership of some of their land, they were pressured to sell at a low price because it had become increasingly difficult for them to continue farming due to decreasing yields in limited spaces: some sold 1.5 ha of their land for just USD50.[4] The sale of land created conflict between villagers because land documents did not exist and property demarcations were not clear. In many cases, the land sold by one villager was being cultivated by another villager and both claimed ownership of the land. Furthermore, when the government confiscated the land, it did not do anything with it, and therefore, no employment was created.

When the villagers of K'dam were upland farmers, they were self-sufficient and did not need to borrow money. After the loss of their land, fishing became their main livelihood rather than their secondary livelihood. A total of 90% of the villagers were full-time fishers in the village in which the head of K'dam Community Fishery lived.

In the early 2000s, as livelihoods in K'dam village shifted from farming to fishing, the villagers needed to buy fishing equipment on credit. At the same time, increasingly more boats came to fish near their shore, which resulted in many conflicts. Fishing nets set by villagers were destroyed by larger fishing boats, which angered the local fishers. Male fishers went onboard the boats to fight, and sometimes they were shot and killed in the process. They lost their fishing gear and livelihoods, and therefore, their loans remained unpaid, which meant their households fell deeper into debt. When the debts started to accumulate, families turned to migration to find alternative sources of livelihood and sent young people in search of jobs.

Individual fishers found that they were unable to fight against those onboard the large boats, and therefore, they jointly took their case to the provincial authorities in 2003. The provincial officials linked them to an NGO that supported the registration of CFs and allowed them to be responsible for fishing in an area of sea covering 24,300 ha. As members of CFs, they are able to inform relevant authorities about illegal fishing boats, and with the support of an official from the Fisheries Administration (the Department of Fisheries before 2006), they can arrest the fishers onboard the boats. They are able to secure a certain percentage of the fines paid by the arrested fishers. However, since there is only one fisheries office in Koh Kong Province, it is difficult from them to arrest illegal fishers. Fishers also complained that even when they caught illegal fishers, the fishers could bribe the officials to release them.

According to the head of K'dam Community Fishery, the unit had 431 members in 2017, all of whom were full-time fishers. Among them, 268 households had motorized boats that enabled them to fish at sea, while other households used small traps for small fish, shrimps, and crabs or gleaned shellfish and snails in and around the mangrove area. Despite decreases in the fish resources to one-third of what they were in 2011 or 2012, the resources are still abundant. Even without boats, fishers can earn KHR20,000–40,000 (USD5–10) per day throughout the year, which is higher than the minimum wage of garment factory workers in Cambodia.

Ms Theary was originally from Kratie Province. Her husband was a native of K'dam village, and she married into the village. After marriage, she accompanied her husband when he went fishing. In addition, she worked as a daily wage labourer picking crab meat. Before the year 2000, life was difficult for the couple, because the area in which the village is located was remote, so it was difficult to sell fish. However, following the establishment of transport connections in 2000, which afforded good access to market, the couple were able to sell fresh fish easily and no longer needed to work for a daily wage. Ms Theary stopped fishing when her husband died in 2008. She started a money-lending business in order to raise her children, and later, her son started to run his own business. She enjoyed working for the community, and therefore, she continued to be active in the K'dam Community Fishery by patrolling at sea and attending meetings. In 2017 alone, the CF made 25 arrests of illegal fishers. It had seven patrol groups, and the members took turns patrolling the area in four patrol boats, each of which could hold seven people.

Ms Theary had been a member of the K'dam Community Fishery from the time of its establishment in 2003 and was elected as head of the village-level patrol groups.[5] Her husband was an ordinary member but was happy that she had been elected. Ms Theary was later elected as a vice head of the K'dam Community Fishery, which had eight committee members, among them one woman, and she held the position from 2008 to 2013. In 2013, she became the head of K'dam Community Fishery, and in 2016, she became the regional head of CFs in four coastal provinces in Cambodia. The year 2013 was important for K'dam Community Fishery because it started to fight cases against external investors and not just illegal fishers. It fought cases against investors who were trying to use the CF's sea and coastal areas for purposes other than fishing. A few significant incidents that occurred in the K'dam Community Fishery are described in the following sections.

### *Conflict with a seaweed cultivation company*

In 2013, part of the area fished and managed by K'dam Community Fishery was occupied by a private company in order to plant seaweed. The company enclosed some areas and prohibited fishers from fishing there. People from six villages that comprised approximately 1000 households protested at the commune office and then at the district office. They also held a demonstration at the location. In addition, they raised the issue at a national workshop, and they blocked the road so that the company workers could not work. By 2014, the seaweed company closed operations in the marine area, and fishers were able to use that part of the sea again.

### *Land grabbing in mangrove forests*

While the K'dam Community Fishery members were still fighting the seaweed case in 2013 and 2014, a shrimp farming company decided to occupy 63 ha of mangrove forest land under the management of the CF. Ms Theary organized a protest against the company; people from three villages joined the protest, and more than 100 women and slightly fewer than 100 men participated. They requested the Ministry of Environment to allow K'dam Community Fishery to manage the mangrove forest area.

### *Violent conflict with fishers on large boats*

In 2015, there were some violent clashes between the fishers in K'dam village and fishers on large fishing boats. The case went to court, but the village fishers had to go to the city of Sihanouk Ville (approximately 100 km east of Koh Kong Province) because the large boats were registered there and because there was no court in Koh Kong Province. This meant they had to spend much time and money on fighting the case. Ms Theary also put time and effort into pursuing the court case. The case was fought by six male members of K'dam Community Fishery, one fisheries official, and Ms Theary, who thought that they had strong evidence. The case has since been suspended.

### *Female leadership in K'dam Community Fishery*

Women were active in both the seaweed cultivation company case and the land grabbing case because they needed to secure their food sources. Relatively few women accompany the men on boats and then only for fishing trips that last under one day. Rather, the shallow water areas and especially the mangrove areas are their 'food pots', where they glean and set traps. According to Ms Theary,

> Men fish, but women are the ones who have to make ends meet by selling fish and managing the household finances. If there is no money,

> women are the ones who have to go to borrow money. Women participate in tree planting, since they depend on these mangrove forests. So, they will bring their children to participate in tree planting.[6]

The men in the village supported the women's participation in their protests, especially after the successful fight against the seaweed company. According to Ms Theary, the men entrusted the women with taking the lead in the protests. However, their support was only for participation in demonstration and protests because they only lasted a matter of days. Women still had difficulty in securing their husbands agreeing to them working as committee members in K'dam Community Fishery.

Ms Theary attributed the success of the protests to women's participation and women's leadership. She said that women had a role in softening the conflict:

> Women are more daring. It is also OK for women to make some mistakes. It is more accepted and they don't mind. Women are the mother of the world, so they [the companies] would not do any harm to women.

The K'dam Community Fishery built a small restaurant near the mangrove forest to start ecotourism in 2017, from which the income went to individual service providers as well as to the CF. The members of the CF run organized groups for boat trips and serve food. They are also preparing to welcome tourists to stay in their homes. So far, they have hosted some student groups and government officials at their ecotourism project site.

As the K'dam Community Fishery's activities attracted more support and interest, village leaders who were pro-investment sought to increase the benefits by allowing private enterprises to develop the coastal area rather than protecting fishing and the coastal resources. When the CF became successful, Ms Theary was ousted as the head of K'dam Community Fishery at end of 2018. Since then, an all-male CF committee has been in place, which is said to be more supportive of the private enterprise investments. Ms Theary subsequently moved an ecotourism project to her home village, where she has remained the head of the village subgroup of K'dam Community Fishery.

Women have played a large role in the development of K'dam Community Fishery, especially when they have been able to devote their time to the CF's activities. They have been able to mobilize other women to diversify the activities of the CF, in addition to patrolling the area. Mangrove forest protection is important for women, especially for women who are poor and do not have boats for fishing with their husbands and therefore have to depend on the resources in the mangroves. One of the problems of CF regulations in Cambodia is that they tend to focus on water bodies and fish resources, and the rights of CFs over coastal areas, including mangrove forests and adjacent land developments, are not guaranteed. The rights are more important for women and for households without boats. Although women have led demonstrations, they still do not receive the recognition that they deserve, as seen in the case of Ms Theary, who devoted 10 years of hard work to become the head of the K'dam Community Fishery. She said, 'Men do not value women's work.'

## The case of Konkang Community Fishery

Konkang Community Fishery in Kampot Province was established in 2008, with 734 members from three villages. The CF has received a lot of support from various organizations, including ActionAid and the EU. The initial activities of Konkang Community Fishery were to protect fishing area by organizing patrols. A total of 20–30 of its members patrolled the CF four times per month. Later, they started to plant mangrove trees. When Konkang Community Fishery was established, it had 100 ha of mangrove forest, which has since been expanded by 30 ha of newly planted area. As in the case of K'dam village, the sea in the area under the management of K'dam Konkang Community Fishery is shallow, and many fishers do not use a boat but instead fish on foot for crabs and shrimps. Konkang Community Fishery has become well known for its ecotourism, which was started in 2014. Approximately 70 people (50 of them women) are involved in the ecotourism business and are engaged in providing accommodation services, restaurants, boat trips, and walkways inside the mangrove forest.

Before starting ecotourism, the members of Konkang Community Fishery fought for their rights to the coastal land in Kampot Province and the coastal waters. In early 2007, a private investor tried to obtain part of the CF's land for hotel construction. Of the 5000 ha of land subsequently required for hotels, 248 ha fell inside the CF's area. The head of Konkang Community Fishery requested a meeting with the provincial governor to regain the rights to the land. It took one and a half years before a meeting was held with the governor, and the meeting was held at night, which made the head of Konkang Community Fishery suspicious and afraid. Despite the delayed opportunity to meet the governor and the timing of the meeting, the negotiation between the governor and the head of the CF, along with another male member who had accompanied him, led to the CF regaining rights to the land.

In 2016, a Thai company planned to invest in the same land. The head of Konkang Community Fishery asked for disclosure of information about the investment and went to court to fight a case against it. The head was offered cash bribes and political power (as a commune chief) in return for agreeing to the investment, both of which he declined. He also enjoyed support from some provincial officials in the initial stage of the court case. However, later, one of those provincial officials was transferred to different position and was replaced by a pro-investment official. The CF head used social networking (Facebook) as a way to disseminate information about the CF and to protect himself and the CF from excess harassment by investors. He raised the issue of land grabbing on Konkang Community Fishery's Facebook page, and he believed that such publicity had protected him and subdued interest in the Thai company's investment: 'Community fisheries only have rights but no power. . . . Rights of community fisheries are guaranteed in law, but we do not have the power to exercise the rights.'

In 2018, a Chinese investor planned to invest in the fishing area managed by Konkang Community Fishery. The case was still ongoing when our study was conducted, and Konkang Community Fishery was finding it difficult to fight against the investment. When the Thai company wanted to grab the land to which the CF had rights, the members of the CF (both men and women, but more women) came together to collect thumbprint signatures in protest against the sale and to protect the area where they searched for food. They also contacted the media. Fisheries Administration officials came from Phnom Penh to discuss the case. Ultimately, Konkang Community Fishery kept its rights to the land back by promising that it would plant mangroves in the area.

Many of the members of Konkang Community Fishery are Muslims, and the degree of social cohesion among the members is high. Although the head of the CF is a man, the level of women's participation in the CF's activities is higher than that of the men's participation. Initially, more men participated, but as the CF developed and land conflicts increased, the women realized that protecting the water body and mangrove forest was important for their livelihoods. Furthermore, the men needed to fish and could not spare time to fight against investors or engage in activities such as mangrove planting and ecotourism. Therefore, the male fishers asked their wives to represent them at the Konkang Community Fishery's meetings, and as a result, more women started to be actively involved in the CF. During the focus group discussions held with women, the participants said the following:

> Before, people said that women work around the fireplace. Now, others work around women and women are the centre of action. Women lead the way.
>
> Before, women were cheap. Now, they are not. They have value.
>
> After Pol Pot, people were scared. Women also did not dare to speak, but now women are the ones who pave the road to lead others.
>
> Before, women were afraid of men. Nowadays, men are afraid of women.

Although women had initially got involved in Konkang Community Fishery because their husbands needed to fish, they have since become the main drivers in CF's ecotourism project.

## CFs' changing role and the emergence of women's leadership

The two case studies show how women can create spaces of leadership when the space that they find comfortable and of which they have a sense of ownership becomes the centre of attention (in this case, the coastal mangrove area where women usually gleaned). In our study, we found that women's leadership is not only triggered by dwindling resources, but also because the marginal space that women occupied became the place of contestation. As Cornwall (2003) notes, creating a space for women to participate in is the first step to enabling women to take up leadership positions. In the studied cases, the attention and threat to the formerly marginal spaces of community fisheries that women occupied led them to take on leadership roles.

The two CFs are both in coastal areas, and most of their members are full-time fishers. None of the women fished alone. Many women gleaned crabs, shrimps, mussels, snails, and small fish in the mangrove forests, both for home consumption and for sale. Some went to the locations by boat, and some went on foot. There are few processing activities in the villages.

In both cases, the CFs had been established following conflicts over fish resources. The CFs' first priority was to protect their fishing resources from illegal fishers. Subsequently, the members of the CFs started planting mangroves to improve the fishing resources. Since fishing from boats offshore is considered men's activity, men were initially more active in the CFs.

Since fishing conflicts started to subside as a consequence of better regulation and since the fish resources have dwindled, conflicts over coastal land and shallow water bodies have emerged. Mangroves and other coastal shallow water areas where women glean are now at risk of exploitation for various purposes, such as tourism and industry. Thus, the contested resources have shifted to the coastal area, where women's livelihoods are based.

The change in focus led to the emergence of women's roles in CFs in the two study CFs. Previously, the work done by the CFs involved mainly patrolling at sea to prevent illegal fishers, but recently their activities have also included negotiating with government authorities and companies to protect the mangroves and coastal areas. We also note that only fishing activities are clearly mentioned in the subdecree relating to CFs, and therefore, the CFs' rights to the coastal land areas can be contested. This recent development and this shift in focus mean that the CFs' focus has shifted from fishing itself to coastal land, which used to be a marginal space for CFs. The CFs used to focus on fishing resources, and coastal resources were largely for gleaning, which was not the main source of income, and therefore, coastal lands were neglected. They were also the spaces where women were most active. Since the focus has shifted to the coastal land, it has created an opportunity for women to have a more visible role. Previously, women stayed at home and gleaned on the coast and in the peripheral areas of the CFs, while the men fished at sea and were at the centre of CFs' activities. Since the coast, as a formerly marginal space, has become highly contested, the situation has created a space where women can exercise their leadership.

The way women's leadership emerged differed in the two studied CFs. In K'dam Community Fishery, a female head of the CF – Ms Theary – emerged when the marginal space that women occupied became the contested space.

Through her dedication and achievements in negotiating with the government and private companies that wanted to invest in the fishing area managed by K'dam Community Fishery, she was able to carve out a space to establish her leadership role in the CF. This was initially accepted by many men, and the majority of Ms Theary's committee members were men. They cooperated with her and acknowledged her leadership, which eventually led to her becoming a regional head of CFs. However, through her work, Ms Theary challenged not only private investors' interests in the coast but also the gender norms that kept women in a secondary role. Ultimately, she was ousted by other men, who took over and shifted the CF in a pro-investment direction.

By contrast, the head of Kongkang Community Fishery was a man. He had been the head since the CF was established. However, female members began to be more active when their struggles relating to coastal land started to emerge, and the women focused on the formerly marginal spaces that they had occupied. However, unlike the head of K'dam Community Fishery, the head of Kongkang continued to serve as a leader, while female members became actively engaged in the ecotourism activities. The female members experienced a high sense of achievement through those activities, but they did not challenge the authority of the male leadership.

The two cases suggest how the emergence of the importance of marginal space creates an opportunity for women to become actively engaged in CFs and to develop their leadership. However, as Meinzen-Dick et al. (2006) note, the organizational environment needs to change so that women's leadership in CFs can be developed further. Unfortunately, in the case of K'dam Community Fishery, the environment and external support was not enough to create an enabling environment for the female head to continue as a CF leader. By contrast, in Kongkang Community Fishery, the male hierarchy was preserved.

Alonso-Poblacion and Siar (2018) argue that diminishing resources are a catalyst for women's participation in CFs. In the two studied cases, not only the diminishing fish catches but also the contest for control over coastal land and mangroves functioned as catalysts for women's leadership. Apart from fighting to protect their coastal land, the change in the CFs' activities from protecting fish resources to diversifying to other income-generating activities, such as ecotourism, also opened up new channels for women to participate in CF activities, which were taken up in both K'dam Community Fishery and Kongkang Community Fishery. However, a deep-rooted notion of leaders being men remained. As seen in the case of K'dam Community Fishery, one woman was ultimately replaced by male leaders, whereas in Kongkang Community Fishery, no women held the position of leader.

## Conclusions

Through a study of two CFs in coastal communities, this chapter has described the rise and fall of women's leadership in CFs. The changes in focus of contestation from fishing area to coastal land encouraged the rise of female leaders, but the persistent male domination in fisheries did not support the women's leadership in the studied CFs. In Cambodia, the CFs are responsible for the management of fishing resources. However, in the studied cases, as the area of conflict shifted from the seas to coastal land and mangrove forests, female leaders rose in the CFs. When the fringe area of CFs was threatened by capital investment from outside, the women at the periphery of the fisheries industry came forward. The women depended on the mangrove forests for food, and they therefore had a strong incentive to protest. The ambivalence in the demarcation of the areas for which CFs are responsible and that earlier had focused only on fish later allowed women some space for leadership, but the space was quickly closed down when men regained power. By analysing the CF's leadership from a gender perspective, we have been able to highlight the unclear definition of CFs' management and rights. It is important that CFs are given management rights to the land around their fishing ground as well as rights not only to fish but also to engage in other income-generating activities for the benefit of their community, such as ecotourism. This will help to recognize and institutionalize women's activities in the fisheries sector, whereas a focus on fishing from boats at sea will highlight only men's roles.

## Notes

1 According to Chapter 2, Article 6, of the sub-decree on community fisheries (Kingdom of Cambodia, 2005),

> A community fisheries . . . is a group of physical persons holding Khmer citizenship who live in or near the fishing area, voluntarily established and taking the initiative to improve their own standard of living by using and processing fisheries resources sustainably to contribute to economic and social improvement and poverty alleviation.

Prospective CFs need to be approved by the Fisheries Administration (in the Ministry of Agriculture, Forestry and Fisheries) in order to be recognized as a CF.

2 The quote, which is an English translation, is from page 36 in an unpublished report titled Pre-feasibility study on project support to the establishment of community fisheries, written by Prak Sereyvath, That Sok, and Lim Sokundarun in 2002 for CEDAC (Cambodian Centre for Study and Development in Agriculture) and JVC Technical College, Phnom Penh.

3 The names of the community fisheries and the study participants are pseudonyms.

4 American dollars are used for large transactions in Cambodia, including those made by villagers.

5 K'dam Community Fishery has members from several villages located around a particular waterbody where people fish. Each village has a subgroup of the K'dam Community Fishery, and each subgroup has representatives on the CF's committee. In most CFs, activities are at the CF level, but in K'dam Community Fishery, the village-level subgroup is active too. K'dam Community Fishery had 18 committee members from various villages, seven of whom were women.

6 All interviews were conducted in Khmer. The translations into English were done by Kyoko Kusakabe, the first author of this chapter, and were checked by members

of the Cambodian Institute for Research and Rural Development (CIRD), who were present during the interviews.

## References

Alonso-Poblacion, E., and Siar, S.V. (2018). *Women's participation and leadership in fisherfolk organizations and collective action in fisheries: A review of evidence on enablers, drivers and barriers*. Rome: Food and Agriculture Organization of the United Nations.

Chap, S., Touch, P., and Diepart, J.C. (2016). *Fisheries reforms and right-based fisheries: Iinsights from community fisheries across Cambodia*. Phnom Penh: Learning Institute.

Cornwall, A. (2003). 'Whose voices? Whose choices? Reflections on gender and participatory development. *World Development*, *31*(8), 1325–1342.

Donnan, H. and Wilson, T.M. (1999). *Borders: Frontiers of identity, nation and state*. Oxford: Berg.

Doss, C.R., and Meinzen-Dick, R. (2015). Collective action within the household: Insights from natural resource management, *World Development*, *74*, 171–183.

Higgitt, R. (2011). Women's leadership building as a poverty reduction strategy: Lessons from Bangladesh. *Journal of South Asian Development*, *6*(1), 93–119.

Hutson, K. (2006). Fishing lot history. *The Phnom Penh Post*, 21 April. Retrieved from www.phnompenhpost.com/national/fishing-lot-history

Khan, S.S., Hopkins, N., Reicher, S., Tewari, S., Srinivasan, N., and Stevenson, C. (2015). How collective participation impacts social identity: A longitudinal study from India. *Political Psychology*, *37*(3), 309–325.

Kingdom of Cambodia. (2005). *Sub-decree on Community Fisheries Management*. Subdecree No. 80/OrNorKror/BorKor. Retrieved from www.ecolex.org/details/legislation/sub-decree-no-80ornorkrorborkor-on-community-fisheries-management-lex-faoc085864/ accessed 5/2/2020.

Kurien, J. (2018). From individual rights to community commons. *SAMUDRA Report*, *80*, 10–13. Retrieved from www.icsf.net/images/samudra/pdf/english/issue_80/270_Samudra_80_Dec_2018_ICSF.pdf accessed 5/2/2020.

Mamun, A. (2015). Leadership in community-based organizations: What fisheries comanagement teaches us? *International Journal of Social Science Research*, *3*(1), 172–188.

Meinzen-Dick, R., Pandolfelli, L., Dohrn, S., and Athens, J. (2006). *Gender and collective action: A conceptual framework for analysis*. Paper prepared for the 11th Biennial Conference of the International Association for the Study of Common Property (IASCP), 19–23 June 2006, Bali, Indonesia. Retrieved from https://pdfs.semanticscholar.org/ddcc/cb0915facd8f7b8aaf319b66b32bb877a413.pdf accessed 5/2/2020.

Meola, C.A. (2013). Navigating gender structure: Women's leadership in a Brazilian participatory conservation project. *Forests, Trees and Livelihoods*, *22*(2), 106–123.

Rowlands, J. (1995). Empowerment examined. *Development in Practice*, *5*(2), 101–107.

Scott, J.C. (2009). *The art of not being governed: An anarchist history of upland Southeast Asia*. London: Yale University Press.

Sheridan, A., Haslam McKenzie, K., and Still, L. (2011). Making visible the 'space of betweenness': Understanding women's limited access to leadership in regional Australia. *Gender, Place and Culture*, *18*(6), 732–748.

Sturgeon, J.C. (2005). *Border landscapes: The politics of Akha land use in China and Thailand*. London: University of Washington Press.

Uchida, Y., Takemura, K., Fukushima, S., Saizen, I., Kawamura, Y., Hitokoto, H., . . . Yoshikawa, S. (2019). Farming cultivates a community-level shared culture through collective activities: Examining contextual effects with multilevel analyses. *Journal of Personality and Social Psychology*, *116*(1), 1–14.

WorldFish. (2013). *Building resilient community fisheries in Cambodia*. Retrieved from http://pubs.iclarm.net/resource_centre/WF_3454.pdf

# 12 Livelihoods, migration, and mobility

## The distribution of consumption expenditure in fishing communities in Cambodia, India, and Sri Lanka

*Ramani Gunatilaka*

### Introduction

An important strand in the economics literature on fishing communities in developing countries concerns their poverty and access to resources, and it makes clear that although fishers are not always the poorest, poverty can be both a cause and a consequence of resource degradation (Allison et al., 2006; Béné, 2003; Béné and Friend, 2011). It has also been argued that technological modernization and demographic growth can lead to 'specific patterns of poverty and riches' in fishing communities (Bavinck, 2011). This chapter adds to the literature by using survey-based data from fishing communities in three Asian countries, to examine the distribution of their consumption expenditure. It focuses on how different livelihoods are associated with different distributions of consumption and explores whether more fishing households that migrate or are mobile in order to access fish resources beyond their immediate environment as a livelihood strategy are better off than fishing households that are not mobile and do not migrate.

I focus on three geographical locations, including inland and coastal areas that may represent different stages in the advancement of the fishing industry in the study countries in terms of stylized facts. In Cambodia, almost half of all households are engaged in small-scale fishing (particularly inland on the Mekong river and the Tonle Sap lake area), although fishing is one of several income-earning activities that households engage in by using traditional methods and fishing gear. Cambodians also migrate for work internally to other parts of the country and migrate overseas, particularly to neighbouring Thailand. By contrast, fishing households along the Indian and Sri Lankan coasts are more specialized, with those in India being the most specialized and using far-more-sophisticated technology and equipment.

In the case of India, the focus in this chapter is on fishing communities in two of Tamil Nadu's coastal districts, Kanyakumari District and Cuddalore

District. Kanyakumari District is on the southern-most tip of the peninsula, on the west coast near Tamil Nadu's border with Kerala. There, fishers engage in deep-sea fishing by using longlines. Cuddalore District is located south of Chennai (the capital of Tamil Nadu) on the east coast, and most fishing households in the district engage in long-distance multiday ring seine fishing along the coast.

The focus on Sri Lanka in this chapter is on the seasonal internal migration of small-scale fishers (most of whom use motorized fibre-reinforced plastic boats[1] or engage in beach seine fishing), from two 'sending' fishing villages in Puttalam District on the west coast to two 'receiving' fishing villages in Trincomalee District on the east coast during the southwest monsoon in Asia.[2] There is conflict over fishing resources between seasonally migrant fishers from Sri Lanka's west coast and fishers in the destination regions on the island's eastern shores. Internal conflict has been exacerbated by external conflict over fishing resources. While fishers from Sri Lanka's west coast have traditionally migrated to the northeast coast to fish, the incursion of bottom trawlers from south Indian into Sri Lanka's northern waters has severely stressed fishing resources, the marine environment, and fishing households that operate there. This has had effects on traditional internal migration, as the Indian incursion has compelled the subnational government of Sri Lanka's Northern Province to actively – even if unconstitutionally – discourage the internal migration of fishers from outside the province. As a result, displaced west coast migrants have crowded onto the coast just south of the northern administrative border, intensifying resource conflict in Trincomalee District. In the following sections, I describe the data and methods used and then present the findings and conclusions.

## Data and methods

The data used for the analysis of migration and household consumption in Cambodia were nationally representative data from the Cambodian Socio-Economic Survey (CSES) conducted in 2011, which were accessed from the National Institute of Statistics (NIS) in Cambodia. The survey covered approximately 3500 fishing and non-fishing households and included a schedule on migration that elicited information about both internal and external migration. For the study of fishing communities in Tamil Nadu, India, primary data were drawn from a survey conducted by the Fisheries Management Resource Centre (FishMarc) in 2017 of approximately 630 fishing households in six villages in the two coastal districts of Kanyakumari and Cuddalore. The data for the Sri Lankan analysis were drawn from a household survey conducted in 2017 of 400 households in two villages in Puttalam District on the west coast and 400 households in two villages in Trincomalee District on the east coast (conducted by the International

Centre for Ethnic Studies [ICES]). The communities in Trincomalee District had been affected by the long military conflict, which ended in 2009, and by the Asian tsunami of 2004. The Indian and Sri Lankan surveys elicited information about the socioeconomic characteristics of households, including demographic characteristics, migration, employment, earnings, consumption, and assets and information about perceptions of subjective well-being, networks, and constraints.

Summary statistics and kernel density functions were used to describe the data in terms of livelihoods, migration, and consumption distributions. Kernel density estimation techniques produce 'histograms'[3] of the distribution of consumption without the dependence on end points of 'bins' and the discontinuity associated with histograms. Since the focus was on the impact of movement on the distribution of consumption in fishing communities, kernel density functions were estimated separately for migrant and non-migrant fishing households. However, because the kernel estimator has difficulty in handling densities that have a high degree of asymmetry, such as per capita household consumption, adaptive kernel density estimation was used with the Epanechnikov kernel, which adjusts the bandwidths to the density of the data points and thus avoids the problem of either over-smoothing and greater bias or under-smoothing and greater variability (for details, see Van Kerm, 2003).

The analysis focused on per capita monthly household consumption expenditure rather than income because it is less likely to be under-reported in household-based surveys. It is also a more accurate measure of money-metric welfare, because households adopt consumption smoothing measures to cope with the variability in income. For the Cambodian analysis, I used the per capita monthly consumption aggregate data constructed by the World Bank for poverty measurement. The consumption data was generated using data from the Cambodia Socio-Economic Survey 2011 and expressed in Phnom Penh prices (Sobrado et al., 2014). Details on the construction are available in the appendices to the report by Sobrado et al. (2014). The consumption data for the Indian villages in Kanyakumari District and Cuddalore District were estimated on the basis of weekly expenditures on food and monthly expenditures on non-food items not related to production activities that households reported to enumerators. It was not necessary to adjust consumption data for spatial differences in prices, because the two districts are in the same Indian state. Consumption data from the Sri Lankan districts included food and other expenditure not related to production activities. Spatial differences in prices between the Puttalam District households and the Trincomalee District households were accounted for by using the district-based poverty lines published by Sri Lanka's Department of Census and Statistics for the relevant survey months to express the consumption data in Puttalam prices (Department of Census and Statistics, n.d.). For the comparative analysis of the distribution of consumption in India and Sri Lanka, the consumption data were adjusted using national price indices (India: the All India Consumer Price Index and the Tamil Nadu Consumer Price Index (Government of India, 2015); Sri Lanka: Colombo Consumers' Price Index (Central Bank of Sri Lanka, 2019) and the purchasing power parity (PPP) conversion factors from the World Inequality Database.[4]

## Livelihoods in fishing communities

### *Overview*

The livelihoods of households in the three country locations are presented in Table 12.1. The data from Cambodia are representative of the entire country, whereas the data from India and Sri Lanka are representative only of the study villages in the respective districts. The livelihood categories have

*Table 12.1* Households' livelihoods in Cambodia (2011) and in studied fishing villages in study districts, both in Tamil Nadu, India (2017), and on Sri Lanka's west and east coasts (2017)

| *Household livelihood* | *Cambodia (%)* | *India: Tamil Nadu* | | *Sri Lanka: west and east coasts* | |
|---|---|---|---|---|---|
| | | *Kanyakumari District (%)* | *Cuddalore District (%)* | *Puttalam District (%)* | *Trincomalee District (%)* |
| Fishing and/or allied fishing only | 0 | 80 | 76 | 37 | 41 |
| Fishing and/or allied fishing and agriculture only | 25 | 0 | 1 | 0 | 2 |
| Fishing and/or allied fishing and non-agriculture only | 0 | 12 | 11 | 31 | 15 |
| Fishing and/or allied fishing, agriculture, and non-agriculture only | 29 | 0 | 0 | 0 | 1 |
| Agriculture only | 7 | 0 | 1 | 0 | 5 |
| Non-agriculture only | 15 | 2 | 6 | 22 | 27 |
| Agriculture and non-agriculture only | 6 | 0 | 0 | 1 | 4 |

(*Continued*)

*Table 12.1* (Continued)

| *Household livelihood* | *Cambodia (%)* | *India: Tamil Nadu* | | *Sri Lanka: west and east coasts* | |
|---|---|---|---|---|---|
| | | *Kanyakumari District (%)* | *Cuddalore District (%)* | *Puttalam District (%)* | *Trincomalee District (%)* |
| None of the above | 18 | 0 | 0 | 10 | 5 |
| Total (%) | 100 | 100 | 100 | 100 | 100 |

Source: Estimated by using data from Cambodia – Cambodia Socio-Economic Survey 2011; Tamil Nadu – FishMarc's survey of migration and livelihoods in fishing communities conducted in 2017; and Sri Lanka – ICES' Survey on Migration and Livelihoods in Fishing Communities, conducted in 2017

Notes: The total number of households in Cambodia was 240,240; in Kanyakumari and Cuddalore 314 and 319 respectively; and, in Puttalam and Trincomalee 400 and 399 respectively. Sample weights used for Cambodia only; statistics for non-agriculture in Cambodia include labour earnings not related to either the fishing or agriculture sectors; the number of households in fishing in Cambodia does not include manufacturing or service-related activities associated with fishing

been constructed to be comparable across the three geographical locations, but there is much more diversity, as shown in Figures 12.1, 12.2, and 12.3.

In Cambodia, 25% of all households combined fishing with agriculture, and 29% combined fishing, agriculture, and non-agricultural work. By contrast, 92% and 88% of households in the Indian districts of Kanyakumari and Cuddalore, respectively, were engaged in fishing or allied activities, while in the Sri Lankan fishing communities in Puttalam District and Trincomalee District, 68% and 59%, respectively, were engaged in fishing and related manufacturing and trading activities. Thus, the different levels of specialization in fishing in the three locations suggest that fishing households in Cambodia represent an early stage in specialization and technological advancement, and fishing communities in India represent a mature stage. The villages in Puttalam and Trincomalee appear similar to the villages in Tamil Nadu in that they appear to have higher concentrations of fishing households and are more dependent on fishing and related activities for household income. Furthermore, the qualitative data suggest that fishing in Trincomalee has undergone rapid transformation since the tsunami of 2004 and the war ended in 2009, given that the fishing population has expanded and fishers have adopted more technologically advanced methods (see Chapter 5, this volume).

Figure 12.1 shows the diversity of Cambodian households' livelihood portfolios in as much detail as permitted by the data. This can be contrasted with Figure 12.2, which shows livelihoods in terms of sector of employment in the Indian fishing communities. A high degree of specialization is apparent. By contrast, a lower level of specialization in fishing and allied activities is prevalent in the Sri Lankan field locations represented in Figure 12.3.

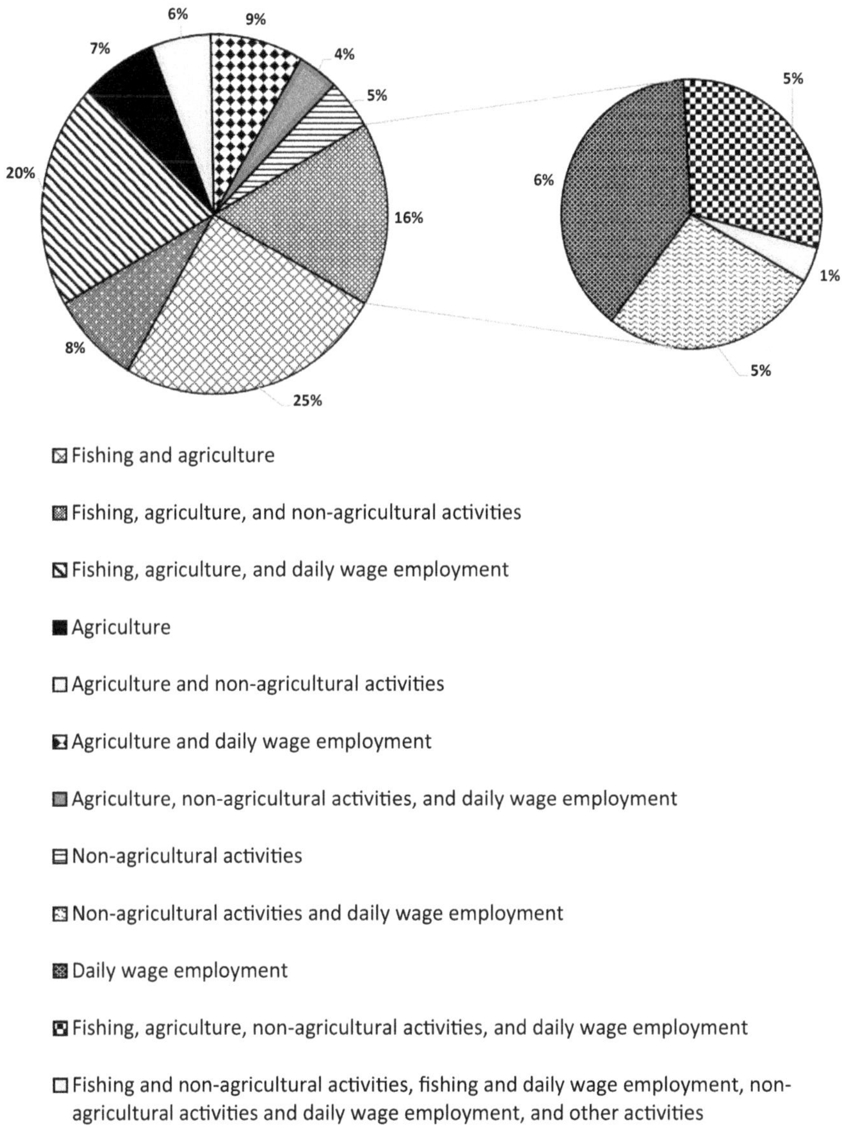


*Figure 12.1* Livelihoods of Cambodian households, 2011

Source and notes: Estimated using Cambodia Socio-Economic Survey CSES 2011. Sample weights used

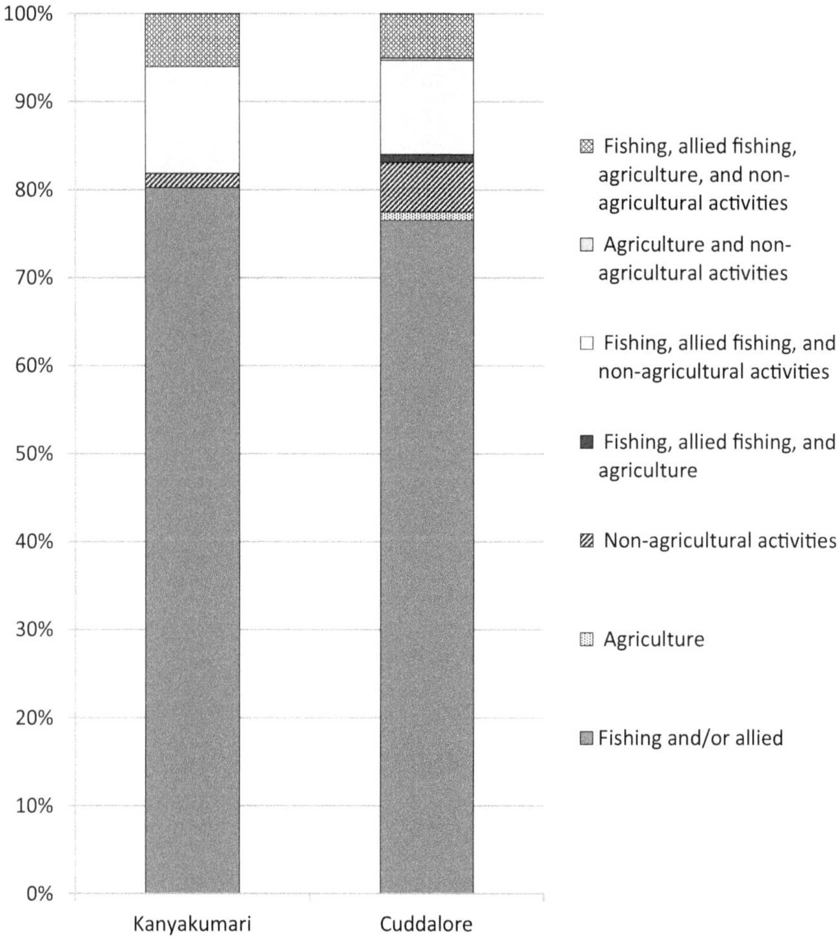


*Figure 12.2* Livelihoods of households in fishing communities in Tamil Nadu, 2017

Source and notes: Estimated using Fishmarc's Survey on Migration and Livelihoods in Fishing Communities 2017.

### *Livelihoods and the distribution of consumption*

Measures of inequality in the distribution of consumption in fishing communities in the three geographical locations are summarized in Table 12.2. The Gini coefficients suggest that inequality was highest in the Puttalam fishing villages and lowest among fishing households in Cambodia. However, the Cambodian data are much older, and inequality in the country may have increased since then.

Consumption inequality in the Indian households was remarkably similar, whereas there was a noticeable difference in inequality as measured by the

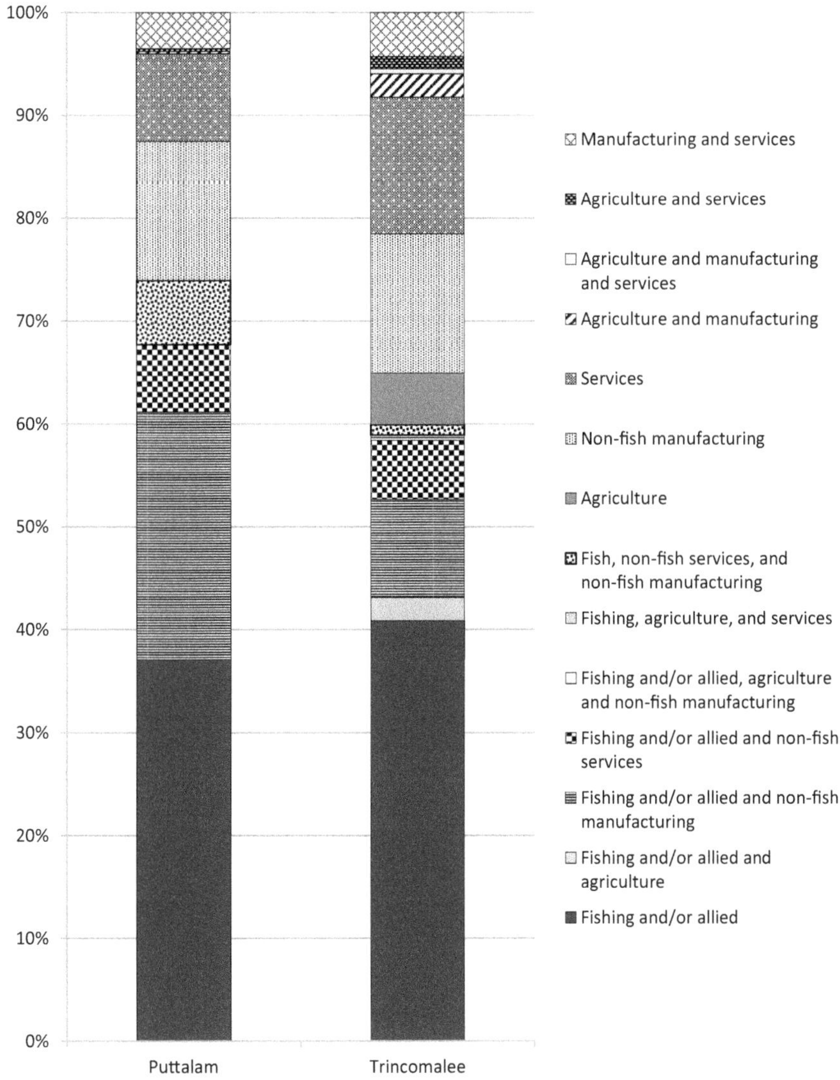


*Figure 12.3* Livelihoods of households in fishing communities on the west and east coasts of Sri Lanka, 2017

Source and notes: Estimated using the ICES' Survey on Migration and Livelihoods in Fishing Communities 2017.

Gini coefficients in the two Sri Lankan districts. In Puttalam, the monthly consumption of households in the 90th percentile was five times as much as the monthly consumption of households in the 10th percentile, while in Trincomalee, those in the 90th percentile consumed three times as much as those in the 10th percentile. However, the summary measures of inequality presented

*Table 12.2* Measures of consumption inequality in Cambodia (2011) and in studied fishing households in Tamil Nadu, India (2017), and Sri Lanka (2017)

| | | | | |
|---|---|---|---|---|
| **Cambodia** | *p90/p10* | *p90/p50* | *p10/p50* | *p75/p25* |
| All households | 3.502 | 2.011 | 0.574 | 1.874 |
| Fishing households | 2.639 | 1.643 | 0.623 | 1.637 |
| | *GE(0)* | *GE(1)* | *GE(2)* | *Gini* |
| All households | 0.220 | 0.240 | 0.333 | 0.367 |
| Fishing households | 0.074 | 0.074 | 0.081 | 0.214 |
| **Tamil Nadu, India: studied fishing households** | *p90/p10* | *p90/p50* | *p10/p50* | *p75/p25* |
| Kanyakumari District | 3.817 | 2.037 | 0.534 | 2.010 |
| Cuddalore District | 5.767 | 1.755 | 0.304 | 2.880 |
| | *GE(0)* | *GE(1)* | *GE(2)* | *Gini* |
| Kanyakumari District | 0.158 | 0.169 | 0.214 | 0.312 |
| Cuddalore District | 0.206 | 0.178 | 0.191 | 0.326 |
| **Sri Lanka's west and east coasts: studied fishing households** | *p90/p10* | *p90/p50* | *p10/p50* | *p75/p25* |
| Puttalam District | 5.402 | 2.617 | 0.484 | 2.448 |
| Trincomalee District | 3.488 | 1.911 | 0.548 | 1.837 |
| | *GE(0)* | *GE(1)* | *GE(2)* | *Gini* |
| Puttalam District | 0.315 | 0.412 | 0.971 | 0.430 |
| Trincomalee District | 0.169 | 0.198 | 0.314 | 0.313 |

Source and notes: Estimated using Jenkins' ineqdeco.ado (Jenkins, 2008), with data from Cambodia – Cambodia Socio-Economic Survey 2011; Tamil Nadu – FishMarc's survey of migration and livelihoods in fishing communities conducted in 2017; and Sri Lanka – ICES's Survey on Migration and Livelihoods in Fishing Communities, conducted in 2017

in Table 12.2 can provide only a partial view of the underlying distribution. By contrast, kernel density estimation methods enable the distribution's level, modality, and spread to be observed simultaneously. Therefore, these techniques were used to visualize the entire distribution of per capita monthly household consumption for each of the geographical locations. Since the focus was on fishing households, the distribution of consumption was estimated separately for some of the fishing-related livelihoods listed in Table 12.1.

In Cambodia, proportionately more fishing households appeared to be less well off than households engaged in livelihoods that did not involve fishing. This is evident in Figure 12.4, which shows the empirical consumption density functions of six categories of Cambodian households according to their livelihoods in 2011, of which three involved earnings generated from fishing. The modal peaks of the latter three are stacked at the lower end of the range of per capita monthly consumption. This shows that most households were among the poorer segments of the population, with households engaged in fishing, agriculture, and labour earnings having the highest proportion of households concentrated along a relatively narrow segment of the lower end of the consumption scale. The modal peak of households engaged in fishing, agriculture, and non-agriculture work is somewhat wider, and

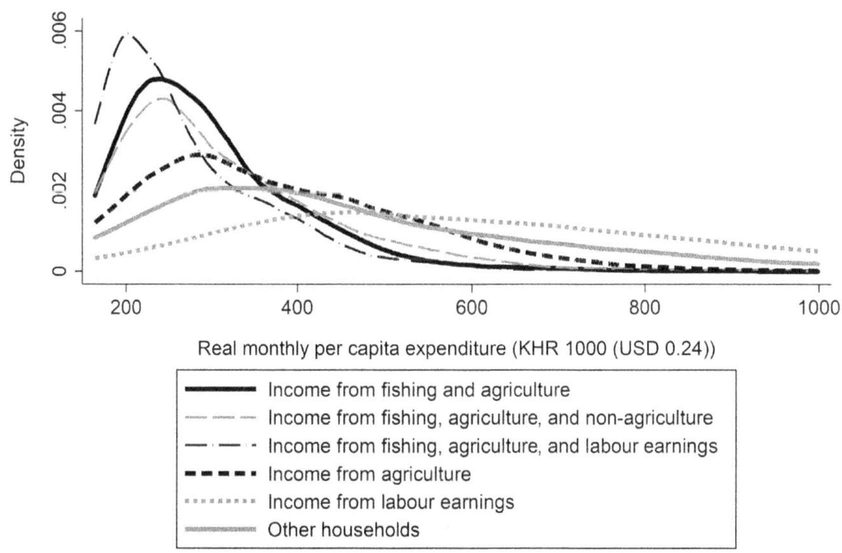


*Figure 12.4* Adaptive kernel density estimation of the distribution of real monthly per capita consumption expenditure by earnings sector, Cambodia, 2011

Source and notes: Estimated by using Cambodia Socio-Economic Survey CSES 2011; sample weights used; and households with per capita consumption expenditure exceeding KHR1 million are excluded from the graph so that the density curves at lower levels of consumption are clearer

the spread of households engaged in fishing and agriculture is widest of all the fishing households. Figure 12.4 shows proportionately more households that engaged in fishing, agriculture, and non-agriculture along the middle segments of the consumption range than households that engaged only in fishing and agriculture. By contrast, the right tails of the empirical density functions of those households that did not engage in fishing are above those of households that included fishing in their livelihood portfolios, suggesting that proportionately more households in those categories are concentrated in the middle ranges of the distribution than are the proportion of fishing households. The results of analysis of household consumption in the fishing villages in Tamil Nadu in 2017 provide an interesting contrast (Figure 12.5). Two features stand out:

1 There are proportionately more fishing and non-fishing households with per capita household consumption along the middle consumption ranges in Kanyakumari than in Cuddalore.
2 More fishing households in Cuddalore were better off than non-fishing households, but in Kanyakumari, more non-fishing households were better off than fishing households.

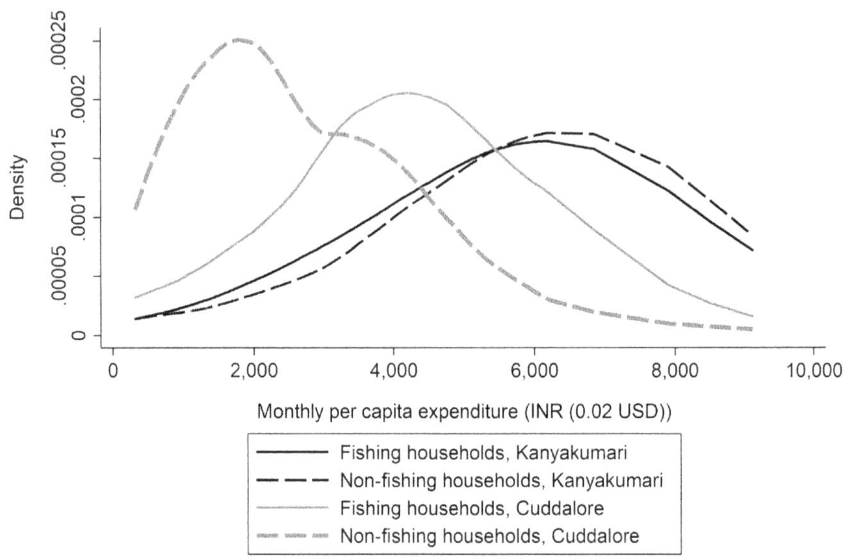


*Figure 12.5* Adaptive kernel density estimation of the distribution of monthly per capita consumption expenditure by migration, fishing households in Tamil Nadu, India, 2017

Source and notes: Estimated by using Fishmarc's Survey on Migration and Livelihoods in Fishing Communities 2017

In Cuddalore, the modal peak of the consumption distribution of fishing households is both squashed down and farther along the range of consumption than the modal peak of the consumption distribution of non-fishing households. By contrast, the right tail of the distribution of non-fishing households in Kanyakumari is slightly above that of fishing households.

In Sri Lanka in 2017, the villages in Puttalam appear to have been relatively more prosperous than the villages in Trincomalee, although the differences between the two districts are not as stark as the differences between the distributions in Kanyakumari and Cuddalore (Figure 12.6). Figure 12.6 shows that in Puttalam there are proportionately more households, fishing and non-fishing, concentrated along the middle consumption ranges of LKR18,000–30,000 (USD120–200) than in Trincomalee. Trincomalee District is on the east coast and much farther away from economically advanced Western Province, which surrounds the metropolitan hub of Colombo. Furthermore, the district was badly affected by the long military conflict. By contrast, Puttalam is on the west coast, is more accessible from Colombo, and was much less affected by the military conflict. In Figure 12.6, fishing households in Trincomalee appear to be in higher concentrations in the lower segments of the range of consumption. By contrast, in Puttalam there are proportionately fewer fishing households in the middle consumption

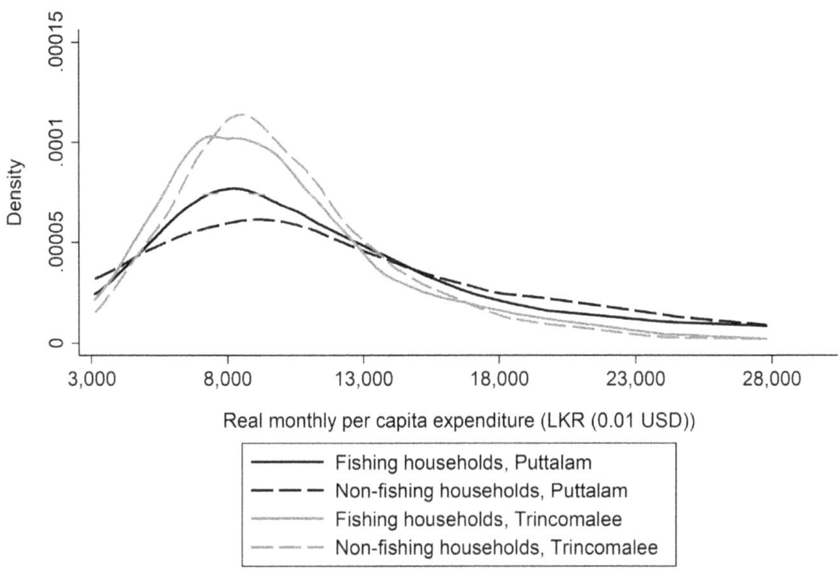


*Figure 12.6* Adaptive kernel density estimation of the distribution of real monthly per capita consumption expenditure fishing and non-fishing households, Puttalam and Trincomalee districts, Sri Lanka, 2017

Source and notes: Estimated by using the ICES's Survey on Migration and Livelihoods in Fishing Communities 2017; households with per capita consumption expenditure exceeding Rs. 30,000 are excluded from the graph so that the density curves at lower levels of consumption are clearer

ranges than non-fishing households, and the modal peak of the curve for non-fishing households is also squashed and spread slightly more towards the middle of the consumption spectrum.

Thus, in the Indian fishing communities in Cuddalore, more fishing households are better off than non-fishing households. In Kanyakumari slightly more non-fishing households are better off than fishing households. However, in Puttalam, there are slightly higher concentrations of non-fishing households than fishing households along the middle consumption ranges. By contrast, in Trincomalee more fishing households are clearly concentrated along the lower end of the consumption scale than are non-fishing households. In Cambodia too, fishing households are clearly concentrated along the lower end of the consumption scale. These results appear to be in line with classical economic theory in that they suggest that greater specialization (in this case, in fishing) is associated with higher productivity and higher returns. In the next section, I unbundle the empirical consumption densities for fishing households further in order to compare the distributions of migrant fishing households with those of non-migrant fishing households.

## Migration, mobility, and the distribution of consumption in fishing households

In Cambodia as a whole, 5% of households had at least one member currently present who had migrated for work during the five years preceding the survey (Table 12.3). Almost twice as many households had members who were currently working elsewhere: 9% in Cambodia and 3% abroad. Households with income only from agriculture appeared to be engaging more in internal migration (almost 25%) than did any other type of household, probably as a risk-mitigating strategy. While there did not seem to be much difference between the proportion of fishing and non-fishing households whose members were currently working abroad, the households that

*Table 12.3* External and internal migration for work by livelihood strategy in Cambodia, 2011

| *Livelihood strategy* | *Household engaged in external migration before 2011 (%)* | *Household engaged in internal migration (%)* | *Household engaged in external migration (%)* | *Household received remittances (%)* |
|---|---|---|---|---|
| Household income only from fishing and agriculture | 3 | 12 | 3 | 14 |
| Household income only from fishing, agriculture, and non-agriculture work | 4 | 10 | 2 | 10 |
| Household income only from fishing, agriculture, and labour earnings | 10 | 5 | 3 | 6 |
| Household income only from agriculture | 0 | 19 | 7 | 20 |
| Household income only from labour earnings | 2 | 3 | 1 | 2 |
| Other households | 4 | 8 | 2 | 7 |
| All Cambodia | 5 | 9 | 3 | 8 |
| All Cambodian households (number) | 143,119 | 274,360 | 83,124 | 240,561 |

Source: Estimated by using data from the Cambodia Socio-Economic Survey 2011

were deriving income from fishing and agriculture seemed to be engaging more in migration in locations elsewhere in Cambodia compared with both other types of fishing households and non-fishing households.

The kernel density functions shown in Figure 12.7 suggest that migration in Cambodian fishing households might have been driven by poverty and that migration might have been a pathway out of poverty. However, there is little to suggest that migration may enable fishing households to move into the middle classes. This is evident in the concentration and spread of the respective modal peaks of the empirical density functions in Figure 12.7. Proportionately more households that had engaged in migration in the past (i.e. before 2011) are concentrated in a narrow segment of the bottom of the consumption scale, whereas the modal peaks of the density functions of those households engaged in external or internal migration at the time of the survey are squashed down and pushed to the right. Furthermore, their right tails are above the right tails of the density function of households that migrated abroad in the past. However, there appear to be two segments in the group of households that engaged in migration in the past, with a small segment concentrated at a range of consumption distinctly higher than the other segment. Thus, some households in which members had migrated in the past but did not migrate in 2011 appear to have done better than others. However, due to the paucity of observations, it was not possible to

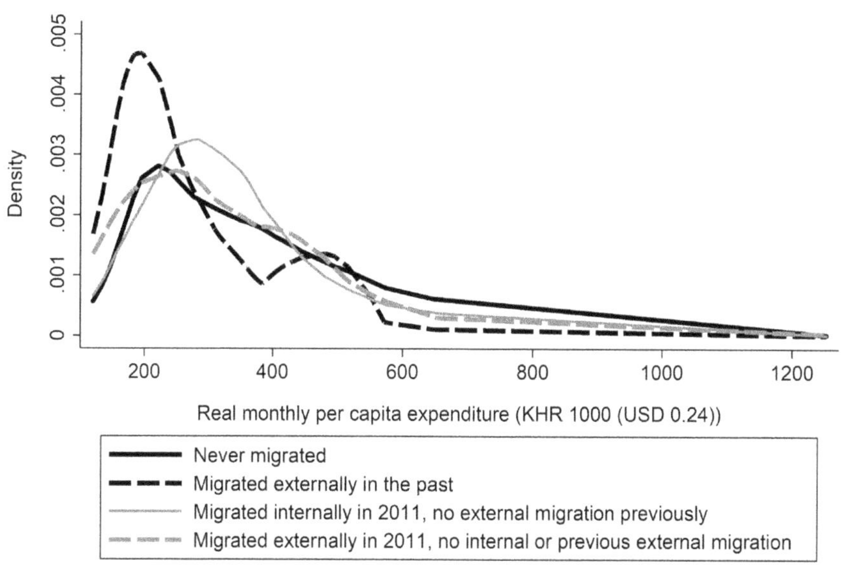


*Figure 12.7* Adaptive kernel density estimation of the distribution of real monthly per capita consumption expenditure by type of migration, fishing households in Cambodia 2011

Source and notes: Estimated by using Cambodia Socio-Economic Survey CSES 2011; sample weights used

investigate the reasons for the result (i.e. to establish whether it was due to the duration of migration or to the reasons for the migrants' return). Nevertheless, the population subgroup of fishing households that never migrated at any time were better off than all other households: in Figure 12.7, proportionately more of them are spread along a wider range of consumption even at the lower end of the scale, whereas proportionately more of them are concentrated along the middle to higher consumption ranges.

By contrast, while fishing communities in Tamil Nadu were far more specialized (as mentioned in the introduction to this volume), they were also more mobile in terms of employment. In this case, movement, not migration, was the predominant means of accessing employment. Since the forms of movement were limited, they are presented in Table 12.4 as numbers rather than as percentages.

Relatively few fishing households had members who migrated to other countries to work. However, most fishing households in Tamil Nadu engaged in fishing at sea by using mechanized boats: slightly more than half of all households in Cuddalore District and slightly more than two-thirds of all fishing households in Kanyakumari District. The kernel density functions of per capita expenditure of such households shown in Figure 12.8 suggest that households in Kanyakumari that engaged in deep-sea fishing and those in Cuddalore that migrated to fish in multiday boats are all concentrated in the middle to upper ranges of the distribution.

The empirical density functions shown in Figure 12.8 were estimated both for subpopulations of fishing households whose members engaged in either deep-sea fishing or multiday fishing in mechanized boats and did not

*Table 12.4* Labour mobility among studied fishing households in Tamil Nadu, 2017

| *Destination* | *Fishing households that migrate for work* | | *Household members who migrate to fish in households that migrate for work* | |
|---|---|---|---|---|
| | *Kanyakumari District* | *Cuddalore District* | *Kanyakumari District* | *Cuddalore District* |
| Same district | 6 | 0 | 5 | 0 |
| Another district | 1 | 2 | 0 | 0 |
| Kerala | 0 | 0 | 0 | 0 |
| Another Indian state | 0 | 0 | 0 | 0 |
| Gulf countries | 10 | 5 | 8 | 4 |
| Singapore or Malaysia | 0 | 10 | 0 | 0 |
| Another country | 0 | 1 | 0 | 1 |
| Deep-sea or multiday fishing using mechanized boats | 163 | 105 | 163 | 105 |
| All fishing households | 234 | 184 | | |

Source: Estimated by using FishMarc's survey of migration and livelihoods in fishing communities, conducted in 2017

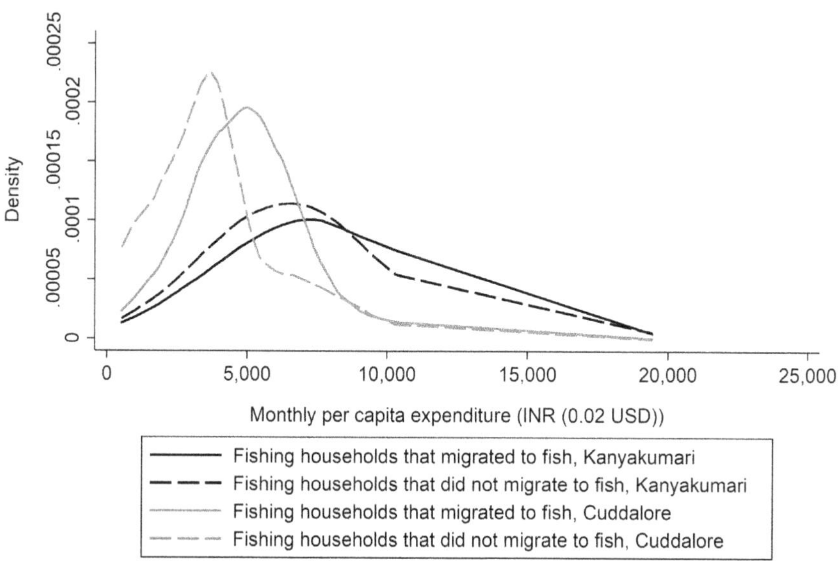


*Figure 12.8* Adaptive kernel density estimation of the distribution of monthly per capita consumption expenditure by migration, fishing households in Tamil Nadu, India, 2017

Source and notes: Estimated by using Fishmarc's Survey on Migration and Livelihoods in Fishing Communities 2017; in Kanyakumari, migration denotes deep-sea fishing that uses long liners, and in Cuddalore, migration involves fishing in multiday boats by using ring seines

migrate to countries such as Singapore, Malaysia, and those in the Persian Gulf and for fishing households whose members did not migrate for work. For both locations (i.e. both Kanyakumari District and Cuddalore District), it can be seen in Figure 12.8 that the density curves of households that migrated are located to the right of fishing households that did not migrate. The right tails of the density functions of households in which members migrated to fish are quite noticeably above the right tails of those that did not migrate. Thus, for both Kanyakumari District and Cuddalore District, households that migrated to fish are concentrated along higher segments of the range of consumption and more so in the case of Kanyakumari than in the case of Cuddalore. This suggests that the practice of migrating to fish is associated with households in the middle classes of the spectrum of consumption.

Information about the numbers of Sri Lankan households that migrated is presented in Table 12.5. The data on migration abroad probably underestimate the full extent of the migration: because the survey was designed to elicit the perceptions of both husbands and wives, both needed to be present at the location where they were interviewed. Therefore, the sample did not include households in which either the husband or the wife was working abroad, even though the qualitative fieldwork suggested that substantial numbers of households belonged to this category. Despite this limitation, the data in Table 12.5 suggest that migration abroad for work was more prevalent in the relatively less prosperous households in Trincomalee than in the households in Puttalam. Regarding internal migration, which was the focus of the study, the data in Table 12.5 show that far more fishing households in Puttalam engaged in the practice than did fishing households in Trincomalee. Therefore, the density functions for only three groups of households in the Puttalam sample are shown in Figure 12.9: fishing households whose members migrated to fish, fishing households whose members do not migrate to fish, and non-fishing households.

In Figure 12.9, the modal peak of the curve for non-migrant fishing households is squashed below the modal peak of the curve for migrant fishing households. However, the right tail of the curve for migrant households is above the right tail of the curve of non-migrant fishing households along the LKR10,000–20,000 (USD67–134) consumption range (i.e. the middle consumption segment). However, proportionately more non-migrant fishing households than migrant fishing households are concentrated along the range beyond the LKR20,000 (USD134) mark. Therefore, the density curve analysis suggests that the migration from Sri Lanka's west coast to east coast in 2017 may concentrate more households in the middle segments of the range of consumption, but not along its upper segment.

*Table 12.5* Household's migration in fishing communities in Sri Lanka, 2017

| *Type of migration* | *District* | *Number* | *Share of total (%)* |
|---|---|---|---|
| All households in which husband or wife had migrated abroad for work | Puttalam | 16 | 4 |
| | Trincomalee | 47 | 12 |
| All households with members living abroad | Puttalam | 25 | 6 |
| | Trincomalee | 7 | 2 |
| Fishing households that migrated internally for employment in fishing | Puttalam | 261 | 75 |
| | Trincomalee | 29 | 11 |
| Total number of households | Puttalam | 400 | |
| | Trincomalee | 400 | |
| Total number of fishing households | Puttalam | 346 | |
| | Trincomalee | 263 | |

Notes: The number of households in the community in which husbands or wives had migrated abroad previously and the number of households with members living abroad probably underestimate the total extent of external migration, given that the survey covered only households in which both husband and wife were living at their normal place of residence at the time of the study.

Source: Estimated by using the ICES's Survey on Migration and Livelihoods in Fishing Communities, conducted in 2017.

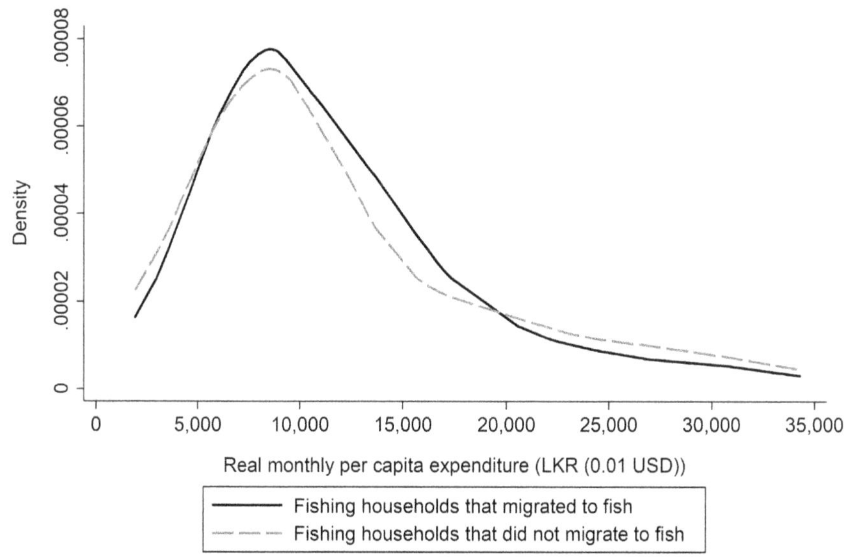


*Figure 12.9* Adaptive kernel density estimation of the distribution of real monthly per capita consumption expenditure and internal migration for fishing, Puttalam, Sri Lanka, 2017

Source and notes: Estimated by using the ICES's Survey on Migration and Livelihoods in Fishing Communities 2017

It was considered of interest to compare the consumption distributions of the fishing communities in India and Sri Lanka. To make the comparison, the per capita expenditure data of the fishing households in the three South Asian locations were converted from their original currency to their to USD equivalents by using the relevant PPP conversion factors and then plotting the adaptive kernel density functions (Figure 12.10).

From Figure 12.10, it is clear that households in Kanyakumari District that engaged in deep-sea fishing were better off than all other households in the study. The left tail of the curve in Figure 12.10 is below all other left tails, showing that Kanyakumari has relatively fewer households concentrated along the lower range of the distribution of per capita consumption. At the same time, the right tail of the density curve is substantially above the right tails of the curves for all the other fishing communities across the entire spectrum of consumption to the right. This shows that Kanyakumari has greater proportions of households concentrated among the higher consumption ranges than all other fishing communities. The right tail of the density curve for migrant households in Cuddalore is above the right tail of the density curve of migrant fishing households in Puttalam but more at the

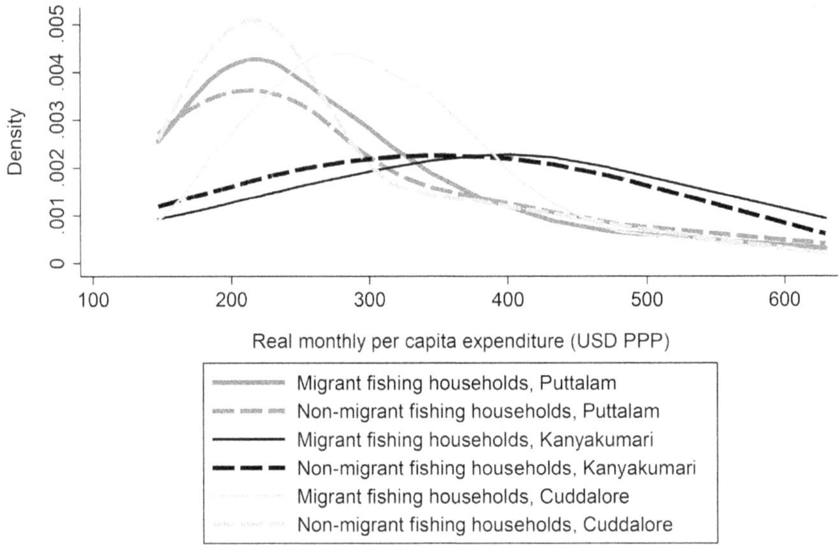


*Figure 12.10* Adaptive kernel density estimation of the distribution of real monthly per capita consumption expenditure of fishing households in Kanyakumari, Cuddalore, and Puttalam, 2017

Source and notes: Estimated by using Van Kerm's (2003) akdensity.ado and data from Fishmarc's Survey on Migration and Livelihoods in Fishing Communities in Tamil Nadu 2017 and from the ICES's Survey on Migration and Livelihoods in Fishing Communities 2017; consumption expenditure data converted USD PPP by using the All India Consumer Price Index, the Tamil Nadu Consumer Price Index, the Colombo Consumers' Price Index, and PPP conversion factors from the World Inequality Database; in Kanyakumari, movement denotes deep-sea fishing that uses long liners, and in Cuddalore, moving involves fishing in multiday boats using ring seines

middle consumption ranges. However, the right tail of the density curve for non-migrant fishing households in Cuddalore is below the right tails of both migrant and non-migrant fishing households in Puttalam along one segment of the middle consumption range. Thus, although Figure 12.10 shows larger concentrations of non-migrant fishing households in Kanyakumari in the higher income ranges compared with migrant fishing households in Puttalam, there are lower concentrations of non-migrant fishing households in Cuddalore along the same ranges compared even with non-migrant households in Puttalam.

Overall, the results of the analysis demonstrated that proportionately fewer fishing households that migrated to or within India and Sri Lanka were concentrated in the lower segments of the income ranges and that proportionately more were concentrated in the higher income ranges than non-migrant households. In the hierarchy of migration or movement, deep-sea fishing households in Kanyakumari were better off than all other households, and multiday fishing households in Cuddalore were the next best off. Although more migrant households in Puttalam were clumped in the middle consumption ranges than were non-migrant households, the difference appeared more muted, and non-migrant fishing households were concentrated in the upper ranges of consumption.

## Conclusions

In this chapter, household survey data from Cambodia, from two districts in India and from two districts Sri Lanka, have been used to understand the relationship between livelihood strategies and the distribution of consumption in fishing communities. In particular, the analysis has looked at how the patterns of consumption distribution of fishing households that practised mobility and migration differed from households that did not migrate for work. Although data from just one survey in each location were used for the analysis and it was not possible to control for the impact of seasonality on the findings, the results of the analysis still provided some key insights into livelihoods, mobility, migration, and the distribution of consumption in the studied communities.

Fishing households in the Indian fishing communities in Cuddalore District and Kanyakumari District were better off than non-fishing households. In Sri Lanka, there appeared to be slightly higher concentrations of non-fishing households in the middle classes in Puttalam District, whereas in Trincomalee District, fishing households were clearly concentrated along the lower end of the consumption scale. In Cambodia too, fishing seems to have been associated with relative poverty, as fishing households were concentrated along the lower end of the consumption scale.

Only a minority of fishing households in Cambodia engaged in migration in 2011, either externally or internally, and they might have been driven to migrate by poverty. Migration might also have been a pathway out of poverty for a minority of the studied households in Cambodia. In Sri Lanka, migration overseas for work was more prevalent in the relatively less prosperous studied households in Trincomalee District than in the studied households in Puttalam District. In Tamil Nadu, India, most fishing households migrated to fish: slightly more than half of all studied households in Cuddalore District used multiday boats to fish along the coast, and slightly more than two-thirds of all fishing households studied in Kanyakumari District used multiday boats for deep-sea fishing. By contrast, three-quarters of fishing households in the Puttalam households migrated internally to fish, compared with only one-tenth of fishing households in Trincomalee.

Greater proportions of the studied households in Kanyakumari District were concentrated among the higher consumption ranges than all other fishing communities. There was also a higher concentration of migrant fishing households in the studied households in Cuddalore District along the middle consumption ranges than there were migrant households among the studied households in Puttalam District. However, although even non-migrant fishing households in Kanyakumari were less concentrated along the lower segments, the non-migrant fishing households in Cuddalore were clumped along the lower segments of the distribution.

Although the analysis revealed that fishing households in Kanyakumari District, Cuddalore District, and Puttalam District that migrated to fish are concentrated in the middle and higher consumption ranges, it cannot be concluded from this evidence that migration or mobility makes fishing households in these districts more prosperous. Compared with the non-migrant fishing households, the migrant fishing households might have had other characteristics or endowments that led to the greater concentration of mobile fishing households along the middle to higher consumption segments in the analysis. Nevertheless, the results of the analysis presented in this chapter reveal where fishing households that migrate and fishing households that do not migrate are placed in the distribution of consumption expenditure. This makes clear which communities and households are better off and may enable policymakers to have a better idea of the distributional impact of policies related to livelihoods, migration, and mobility in the studied communities.

## Notes

1 In Sri Lankan fisheries reports and statistics, the abbreviation OFRP (Outboard-engine Fibreglass Reinforced Plastic) boat is commonly used to refer to small motorized fibre-reinforced plastic boats
2 Batticaloa District and Ampara District were not covered in this component of the study, although the two districts were covered in the study of social exclusion using qualitative data and methods.
3 A histogram summarizes the distribution of observations in terms of a diagram consisting of rectangles. A histogram summarizing the consumption data of a fishing community would be constructed in such a way that each rectangle's area would be proportional (shown on the Y axis) to the number of households whose per capita consumption would fall within the class interval of consumption shown on the X axis.
4 The database is accessible at https://wid.world/data/ accessed 5/2/2020

## References

Allison, E.H., Horemans, B., and Béné, C. (2006). *Vulnerability reduction and social inclusion: Strategies for reducing poverty among small-scale fisherfolk*. Paper presented at the Wetlands, Water and Livelihoods Workshops, Wetland International, St. Lucia, South Africa, 30 January – 2 February.

Bavinck, M. (2011). Wealth, poverty, and immigration: The role of institutions in the fisheries of Tamil Nadu, India. In S. Jentoft and A. Eide (Eds.), *Poverty mosaics: Realities and prospects in small-scale fisheries* (pp. 97–106). Dordrecht: Springer.

Béné, C. (2003). When fishery rhymes with poverty: A first step beyond the old paradigm on poverty in small-scale fisheries. *World Development*, *36*(1), 945–975.

Béné, C., and Friend, R.M. (2011). Poverty in small-scale fisheries: Old issue, new analysis. *Progress in Development Studies*, *11*(2), 119–144.

Central Bank of Sri Lanka. (2019). *Annual report 2018*. Retrieved from www.cbsl.gov.lk/en/publications/economic-and-financial-reports/annual-reports/annual-report-2018 accessed 5/2/2020.

Department of Census and Statistics. (n.d.). *Poverty*. Retrieved from www.statistics.gov.lk/page.asp?page=Poverty accessed 5/2/2020.

Government of India. (2015). [*Untitled*]. *Open Government Data (OGD) Platform India*. Retrieved from https://data.gov.in/catalog/all-india-consumer-price-index-ruralurban?filters%5Bfield_catalog_reference%5D=85827&format=json&offset=0&limit=6&sort%5Bcreated%5D=desc accessed 5/2/2020.

Jenkins, S. (2008). *INEQDECO: Stata module to calculate inequality indices with decomposition by subgroup*. Statistical Software Components S366007. Retrieved from https://ideas.repec.org/c/boc/bocode/s366002.html

Sobrado, C., Neak, S., Ly, S., Aldaz-Carroll, E., Gamberoni, E., Arias-Vazquez, F., . . . de Groot, R. (2014). *Where have all the poor gone? Cambodia poverty assessment 2013*. Washington, DC: World Bank Group. Retrieved from http://documents.worldbank.org/curated/en/824341468017405577/Where-have-all-the-poor-gone-Cambodia-poverty-assessment-2013 accessed 5/2/2020.

Van Kerm, P. (2003). Adaptive kernel density estimation. *Stata Journal*, *3*(2), 148–156.

# 13 Small-scale fishers, mobility, and the politics of well-being in rapidly changing Asia

*Ragnhild Lund*

## Introduction

This book calls for a grounded, multidimensional approach to the analysis of gendered mobilities in small-scale fisheries in Asia. The contributing authors ask how the well-being of small-scale fishing households affect – and are affected by – practices of mobility and social relationships, and they situate these practices in a context of increased market exposure, changing state development initiatives, new social protection, and environmental changes. The individual chapters provide evidence for a more nuanced understanding of how movement and migration among fishers contribute to sustain livelihoods in coastal areas, and they explore how networks of male and female fishers ensure resilience in local fishing communities.

Detailed insights into the ethnographies of individual-country studies show how exclusion and inclusion occur on the ground, creating new patterns of well-being but also vulnerability and precarity. This connection, we argue, is not well covered in research on the processes of differentiation in fishing communities. We find that current vulnerabilities in fishing communities are both old (related to inequalities of caste, class, and gender relations) and new (related to the impact of large-scale modernization and mechanization, the development of new technology, new divisions of labour, and limited access to fishing grounds). Hence, there is an urgent need to understand the possibilities for small-scale fishers in pursuit of a better life, into which we believe that this book offers some insights.

## Lessons learned

These concluding remarks emphasize that although our work draws on detailed ethnography, by going beyond single case studies, we have provided a way of more generally investigating the agency and structures that are constraining small-scale fishers in Asia. The strength of the research on which this book is based is its focus on the nexus between social transformation, multidimensional aspects of well-being, and mobility, with gender at the heart of the nexus. This nexus provides an important context for examining the efficacy of mobility as a strategy for reducing poverty and enhancing well-being among fishing communities in Cambodia, India, and Sri Lanka and for providing an approach that is replicable more broadly for understanding the pathways out of poverty for fishing communities and other marginal groups. The conceptual framework (Chapter 1) draws on decades of feminist interventions that have emphasized the importance of agency for understanding gendered well-being outcomes and how agency is exercised with respect to different institutions (e.g. household, community, market, and state) (Kabeer, 1996, 2007). Furthermore, the framework exposes the exclusionary elements of collective processes, and thus, the contributions in this book are instructive also in the policy realm. Mobility and collective networks are negotiated through gendered access and socially contested claims to resources, recognition, and participation. Thus, the link between agency and well-being is highly mediated.

To recapitulate the main research questions of our study, the aim of the research project was threefold. First, it aimed at unravelling whether migration or movement is a potential route out of poverty for people in fishing communities in the three study countries (Cambodia, India, and Sri Lanka) by identifying disabling conditions (vulnerabilities) and enabling (capabilities) conditions. Second, the study sought to understand the role of social networks and collective action in advocating supportive policies for fishers and their communities to reduce risk, violence, and conflict and to provide insights into their attempts to secure livelihoods and maintain their identities as fisherfolk. Third, the project aimed at uncovering whether current national policies relating to gender, fisheries, and migration are adequate for addressing the vulnerabilities and aspirations of households in fishing communities.

Three major conclusions may be drawn from the country studies. *First*, across countries, while poverty was the main cause of migration for some people and communities, it was not for others. In India and Sri Lanka, migration is driven by a multiplicity of factors and therefore does not necessarily take people out of poverty, although it might help some to stay out of poverty. Particularly in Cambodia, it was found that migration was not a pathway out of poverty but a way to sustain fishing livelihoods. *Second*, the benefits and costs of migration are distributed differentially among women and men, ethnic groups, castes, and classes in the different study locations in the three countries. *Third*, social networks are critical in enabling and disabling migration, but collective action based on these networks is sporadic, and its effectiveness depends on the strength of bridging and linking networks and on the way governments define fishing communities.

The three conclusions are explored further in the individual chapters, which provide additional insights. Chapter 1 reviews the international discourses on small-scale fisheries and presents contextual challenges, as well as the concepts and methodology of the research project on which the book is based. In Part I, the chapters address specific gendered aspects of mobility and migration, ethnicization, and social exclusion. These chapters add knowledge to the literature on the increasing forces of mechanization and capitalization of fishing (Chapter 2), the environmental situation and access to resources (Chapters 3 and 4), and the expropriation of coastal land (Chapter 5).

Furthermore, as documented in Chapter 1, in most of the literature on small-scale fishing, gender has almost entirely been presented as a woman's issue. However, this book documents how both men's and women's situations are changing at all levels in society. For instance, one such change is in how dowry has become a source of capital and status at the individual level, and increasing capitalization has contributed to new forms of organization among male fishers that challenge traditional hierarchies of class and age, triggering also a search for alternative livelihoods (Chapter 2). The authors who present the results of the Indian study (Chapters 2 and 7) address the marginalization of women's roles in fishing in particular, alongside increasing dowries and potentially declining status, as a key finding of their research.

One of the chapters on fishers in Cambodia (Chapter 3) highlights that fishers are not necessarily the poorest people, but they are among the most vulnerable. Given this situation, their expressed preference is to continue fishing. Migration for work and women's non-fishing income are used to support fishing, not replace it. They provide a way to bear vulnerability by allowing fishers to continue fishing without much change in their core occupation. Although women and adult children play a large role in sustaining fishing through their migration and non-fish-related employment, their work is defined as supplementary to fishing, and hence, it does not receive recognition and is not reflected in the gender relations in the household.

In India, much of the existing research has focused on fisheries per se, and the implications of changing technologies and practices, thereby raising the issue of labour relations, and the findings relating to labour mobility are a key contribution of the project (Chapter 2). While fishers and labourers who work as crew on boats are mobile, the nature of their mobility and implications for their well-being and for gender relations have not been explored to any extent. Another key insight from the Indian study is the scale of in-migration from non-fishing areas and communities and the occurrence of migration due to distress. By contrast, the Sri Lankan studies analysed how unequal benefits and costs are perceived by migrant and host fishing communities, how these perceptions underlie emerging resource conflicts, and how potential conflicts are contested through two discourses: a right to continue a tradition of migration and a right to local resources. The study of west (Puttalam) to east (Trincomalee) migration showed that horizontal inequalities among migrant and host fishing communities are exacerbated by vertical inequalities among fishing communities and regulatory institutions (Chapter 4). However, there are regional variations in the country. The study of Batticaloa District and Ampara District in Sri Lanka showed how traditional small-scale fishers have left fishing and migrated overseas due to precarious situations that developed as a consequence of the 2004 tsunami and war (Chapter 5). Female household heads (FHHs) were identified as a vulnerable group and their mobility and migration as intertwined with exclusion at individual and collective levels. Hence, geographical variations and different precarities exist even within a country and across districts.

The quest for alternative livelihoods and well-being among different groups of fishers is evident in Part II of this book. The authors interpret well-being in different ways: whereas some use the social well-being approach specifically, others refer to well-being as social relations and as empowerment. Various chapters show how by using a social well-being analytical lens, we delineate holistically the material, relational, and subjective factors that motivate migration in fishing communities. For instance, the situation in Sri Lanka indicates how these factors shape decisions and trade-offs, and it points to the challenge of reconciling the well-being of women and men and of migrant and host communities (Chapter 6). Relational costs are often higher for women than for men, and compared with migrants, host communities perceive more costs than benefits. In India, pride in the particular identity as deep-sea 'shark hunters' drives male fishers to take considerable risks, yet this is central to their personal well-being and status as skilled fishers, entitling them to respectable social positions (Chapter 7). Women who are left to manage household resources are often in a precarious position, but they draw on their social relationships to ensure household well-being (Chapter 5). Another chapter highlights the plight of those who are not able to migrate, namely ethnic Vietnamese people in Cambodia, whose citizenship is not recognized by either of the states (Chapter 8). Their non-legal status confines them to live on the water. With limited options, households in which the gender division of labour is more flexible are seen as faring better in coping with changes.

More generally, the various chapters in Part II show how well-being is multidimensional and context-specific, shaped by different world views and social conditionings, such as gender, caste, and household composition. Moreover, through unravelling the context-specific content and practices of well-being, the authors document how the differentiated opening up or closing down of opportunities for viable livelihoods, increased competition over fishing grounds, and the local negotiations of national policies all have diverse and complex outcomes.

Part III, on social networks, provides a valuable addition to the study of small-scale fishers. By moving from conceptual understandings towards practical interventions that are based on the ground realities of fishing households, the authors' reflections on networks as different forms of social capital reveal different kinds of interpersonal relationships and how male and female fishers strategize and fight to protect their right to use resources and overcome their vulnerabilities. In India and Sri Lanka (Chapters 9 and 10), the emphasis is on social relations and how they play out in networks.

For instance, while bonding networks are important in both migrant and host fishing communities in the Sri Lankan study locations, stronger livelihood support from both bonding and bridging networks and stronger linking networks among migrant communities enable them to engage in collective action to pursue migration and withstand opposition from host communities. In India too, migration outcomes appear to be more beneficial for fishers in Kanyakumari District, with stronger bonding and bridging capital than that in Cuddalore District, where more instances of distress and failed migration occur. These findings show how bridging and linking ties in fishing communities help to sustain migration and enhance well-being outcomes. In Chapter 11, on Cambodia, the focus is on how female leaders create spaces in a male-dominated institution. The research findings show that the contested area for community fisheries has shifted or expanded from protecting fish resources to claiming coastal land. The change in focus has opened up an opportunity for women to take up leadership roles. However, the male-dominated nature of the organization and lack of external support have meant that women have not been able to maintain their leadership positions.

## Value added

By employing a mixed-method approach, the contributors have drawn the seemingly distant scales of the everyday and political-economic restructurings together to map the effects of agency and institutional structures on the well-being of small-scale fisherfolk. The econometric contribution of the research project allows for cross-country comparisons (Chapter 12), because it documents geographical and economic variations in how migration or movement impacts poverty and the distribution of consumption. Methodological triangulation contributed to the analyses of the various country studies. In the Sri Lankan case, it enabled a nuanced analysis of networking and how migration represents a pursuit for well-being rather than being a consequence of poverty and resource scarcity.

Two other studies were based purely on qualitative methods (Chapters 5 and 8). Their major contributions are through individual stories to provide voices for marginalized populations: ethnic Vietnamese people in Cambodia, who have managed to carve out a niche in the fishing communities, and female household heads in Sri Lanka, who have been displaced due to the 2004 tsunami and civil war but who strategize by migrating abroad. These studies showed how complex the local lives are in terms of multiple gender relations, occupation, and movements, hence providing nuanced pictures of multiple deprivations.

Another key concern of our project was to identify pilot projects and potential strategies for advocacy, and for that purpose, local stakeholder workshops and participatory activities were organized. The pilot projects were identified on the basis of the ongoing research and in partnership with the research participants in the villages themselves. It is beyond the scope of this book to go deeply into the role of pilot projects and dissemination activities, but the chapters on social capital and networks identify possibilities for new practices and actions on the basis of our research. In the Indian case, the focus of the pilot projects was on the most vulnerable, namely the nontraditional Irular fishers and in-migrants from other parts of India, with attempts at developing new practices and instigating social and political action to improve their livelihood security. For the Irulars, support was secured from a non-profit development agency to make a video as part of an advocacy strategy for their livelihood security and designed to reduce both exploitation and conflict. Discussion has also been initiated across sites to provide some recognition of in-migrants, especially those who work as crew in high-risk environments. Debate on the issue of dowry has also been initiated among the fisher communities, although this remains a deep cultural and economic problem in India that must be addressed in a long-term perspective.

In the Sri Lankan case, the pilot projects supported youths to engage in participatory documentary film-making, following the fieldwork period. This provided an independent forum for the youths to voice their perspectives on their livelihoods, migration, and future aspirations and for the researchers to pay attention to the voices of the youths. The youths' perspectives consolidated and complemented the overall study findings and enabled researchers to gain a better understanding of social divisions (gender, caste, and neighbourhood) and dynamics in the study villages.

In Cambodia, the pilot projects supported female leaders in the study area to use the coastal land for generating income for community fisheries through developing community-based tourism in the mangrove forest. Collective action in community fisheries in Cambodia has shown some effectiveness in protecting the fisheries' resources, especially as women mobilize in collective actions. The Cambodian research team recommended revisiting the 2005 subdecree on community fisheries and expanding the designated area of community fisheries to include coastal land and not just fishing grounds. The same team recommended improved employment opportunities and wages for women and youths, and such improvements in labour market conditions should be considered as part of fisheries policy, since without such support, it would not be possible for sustainable fishing to continue in Cambodia.

## Challenges ahead

Our research has revealed recent environmental changes induced by climate change and natural hazards, such as tsunamis and floods, which increased the vulnerabilities of many fishers in the study countries, reduced their access to fisheries resources, caused declines in fish stocks and fish catches, and changed livelihoods. Resource degradation also led to increasing contestations, even conflicts, over fish resources, and in other cases, fishers have moved out of fishing. A major challenge ahead is to find ways of good governance at local, regional, and national levels.

Collective initiatives play important roles for fisheries in all three study countries. However, currently, such initiatives focus mainly on fisheries management and the conservation of fisheries resources, which largely depend on cooperation between different actors. Increased community participation in management could lead to better governance at the local level given that it could create and strengthen spaces for participation. When spaces for community participation are open, communities become aware of the resources that they have and the need to manage, conserve, and develop those resources to alleviate poverty and improve their own well-being.

Furthermore, our research has made visible the need for transformative policies. A review of the transformative potential of relevant policy documents was undertaken in the first phase of the research project and painted a cross-national and regional picture of existing plans and policies with relevance to the fishing sector and fishing populations. The review enabled the researchers to gain a cross-country overview of existing instruments (and the lack of them) to manage, control, and support fishing activities (Azmi, 2018). The formal report based on the review complies with the quest of the United Nations Research Institute for Social Development (UNRISD, 2016), which claims that social structures and relations need to change, shaped by the growing economic power of elites and differences based on class, gender, ethnicity, religion, and location, which constrain peoples' choices and agency. It is important for transformative change to address norms and institutions that shape behaviour and the organization of social, economic, environmental, and political spheres in the long run. Such change requires both individual and collective action. UNRISD further suggests that policies aiming to bring about transformative change should not undermine existing systems, strategies, and institutions (UNRISD, 2016). Thus, transformative polices should not neglect the context in which they operate but instead must reflect an understanding of existing social, economic, institutional, and legal aspects. Transformative change also requires different sectors to work together towards a common objective. Hence, policy coherency, complementarity, and coordination become essential characteristics of transformative change (Azmi, 2018).

The focus of this book is limited to the sustainable development perspective (changes in economy, society, and the environment) and changes in norms and institutions, with attention to poverty and well-being. However, the individual country studies showed how individual policy recommendations must be country-specific, and therefore, the findings were discussed with various country stakeholders. This emphasizes the need to focus more effectively on the conservation and management of fisheries resources through community participation. Women have often been left out of governance structures, at both the local level and the sectoral level. Opening the participation space for different stakeholders provides opportunities to understand emerging needs and priorities and to reorient policies to improve the lives of fishers. The creation of such spaces could also facilitate shared visions and coherent strategies and help institutions to understand and reinforce the links between sectors. By incorporating gender into fisheries sector policies and strengthening existing cooperatives, governments could try to bring about changes in norms and institutions in the long term.

However, our research documents indicate that policies fail to consider the implications of changes in fishing technologies for people as gendered subjects. Shifts to industrial and harbour-based fisheries, such as in the case of India, alongside the takeover of coastal land for tourism and other industrial enterprises, have contributed to a decline in women's access to fish and spaces for post-harvest processing and sale. While women are marginalized from economic opportunities in the fisheries sector, growing male absences from the home, including for multiday fishing and other activities, have increased women's responsibilities for the everyday management of the home. Such insights have significant implications for policy across sectors, relating not only to fishing but also to migration and gender. Currently, in the absence of such policy coherence, the insights' potential for leading to transformative change by reducing poverty and improving well-being is limited and not fully recognized. Furthermore, the theories we have used document how vulnerability and precarity on one hand and mobility and migration on another are contingent on dynamic social change. We find that policies are not flexible enough to respond to some of the dynamic changes that we have unravelled.

Finally, we wish to emphasize that the value of studying and engaging with small-scale fishing is that global processes or networks are not disconnected from the local. As documented in several of the chapters, fishing is increasingly becoming a globalized profession (cf. the discourse on the Blue Revolution). Although our work draws on detailed ethnography, we have provided a way of investigating the agency and structural constraints of small-scale fishers in Asia more generally. In doing so, we have simultaneously related the following to scale: how the global and regional processes of capitalization and mechanization give rise to stress and place limitations on local small-scale fishing but at the same time provide opportunities for other livelihoods. This situation may be understood as a process that combines the concerns of localism with the forces of globalization or a local adaptation and interpretation of global forces, because, as argued by scholars, any external influence has an origin (a consequence of somebody's action), a means of diffusion (e.g. ships or fishing gear), and places or people that 'crystallize' the consequences (Backhaus, 2003; Latour, 1993). In our case, this means that wherever restructuring of fishing, fish processing, and marketing take place, it has effects that will depend on the particular setting or context. Some authors have called this convergence of global dynamics and local contexts glocalization (Backhaus, 2003; Robertson,

1992; Swyngedouw, 1997, 2004). Glocalization expresses the way global dynamics are always reinterpreted locally, leading to an interpenetration of the local and global scales that creates context-dependent outcomes. We acknowledge that such external – internal, local – global interactions exist, and in response, our research approach in this book has been grounded in fishers' lived realities and how they move and strategize in their search for well-being in a rapidly changing society. We consider that our work contributes insights into what constitutes such 'glocal' realities at different levels and thus emphasizes the significance of acknowledging the local–global link in research and the policies related to fishing.

## References


Azmi, F. (2018). *Poverty reduction and well-being in fishing communities in Cambodia, India and Sri Lanka through the policy net*. Unpublished manuscript, Department of Geography, NTNU, Trondheim, Norway.

Backhaus, N. (2003). *Zugänge zur Globalisierung. Konzepte, Prozesse, Visionen*. Schriftenreihe Anthropogeographie 17. Zürich: Geographisches Institut Universität Zürich.

Kabeer, N. (1996). Agency, well-being and inequality: Reflections on the gender dimensions of poverty. *IDS Bulletin*, *27*(1), 11–21.

Kabeer, N. (2007). *'Footloose' female labour: Transnational migration, social protection and citizenship in the Asia region*. IDRC Working Paper on Women's Rights and Citizenship. Retrieved from www.idrc.ca/sites/default/files/sp/Documents%20EN/WRC-WP1-Kabeer-Migration.pdf accessed 5/2/2020.

Latour, B. (1993). *We have never been modern*. Cambridge, MA: Harvard University Press.

Robertson, R. (1992). *Globalization: Social theory and global culture*. London: Sage.

Swyngedouw, E. (1997). Neither global nor local: 'Glocalization' and the politics of scale. In K.R. Cox (Ed.), *Spaces of globalization: Reasserting the power of the local* (pp. 137–166). New York: Guilford Press.

Swyngedouw, E. (2004). Globalisation or 'glocalisation'? Networks, territories and rescaling. *Cambridge Review of International Affairs*, *17*(1), 25–48.

UNRISD. (2016). *Policy innovations for transformative change: Implementing the 2030 agenda for sustainable development*. Retrieved from www.unrisd.org/80256B42004CCC77/(httpInfoFiles)/2D9B6E61A43A7E87C125804F003285F5/$file/Flagship2016_FullReport.pdf accessed 5/2/2020.

# Index

Note: Page numbers in *italic* indicate a figure and page numbers in **bold** indicate a table on the corresponding page.


activities: fishing-related 30, 79, 96, 97; fish-processing 100, 178
adaptation strategies 41, 44, 45, 132, 135, 138–140, 143; gender relations transformation 140–142; household size and structures, determinants 138–139; social network, determinants 139–140
adult children 41, 46, 50, 52, 53, 220
agents 31, 34, 70, 149, 154, 155, 157, 158, 161
agriculture 10, 42, 43, 68, 82, 87, 96, 147, 181, 200, 204, 205, 208, 209
Alexander, M. 47
Alonso-Poblacion, E. 183, 192
Amarasinghe, O. 5
Ampara District 75, 76, 79, 81, 83, 88, 102, 220
An 43
*anbiams* 113, 150, 153, 155, 158
aquaculture 2, 11, 137, 140
artisanal fishers, small-scale 26, 116
Asia: small-scale fisheries 218
Association of Deep Sea Going Artisanal Fishermen (ADSGAF) 115
Aswathy, P. 2
auctioneers 31–33, 122, 147
Ayyankeril, D. 7

Baas, M. 149
Barnett, J. 45
Batticaloa District 64, 67, 75, 76, 79, 88, 165, 170, 220
beach seines 62, 69, 70
Béné, C. 41, 44
Bennett, N.J. 61
Bettini, G. 44–45
Black, R. 44
'blue economy' initiatives 4
Blue Revolution 5, 25, 60
boat owners 11, 12, 34, 36, 37, 84, 87, 103, 122, 154, 156, 157, 168–170; wives of 107, 108
boat ownership 26, 122, 153, 157, 158
boats 12, 31–33, 35, 36, 64, 66, 67, 69, 70, 104, 114–122, 155, 156; carrier 31–32; large 156, 185, 187; main 120; mechanized 114, 210; migrant 69, 70; migratory 70; multiday 13, 210, 215; new 103, 117; overcrowding 65, 67; small motorized fibre-reinforced plastic 96, 167; steel 28, 29, 31, 36; types **29**
bonding, social capital 13, 148, 154, 156–158, 162, 167, 168, 170, 172, 174, 178, 222; networks 168, 170, 175, 178, 222; relationships 153
Bourdieu, P. 9
bridging ties 168–170, 173, 178

cages 139, 140
Cambodia 11, 13, 41–45, 48, 132–144, 181–183, 196–216, 219, 222, 223; community fisheries in 181–194; fishing communities in 42–44; fish resources in 41–54; rural 42, 46
Cambodian citizenship 11, 132, 136
Cambodian fishing households 209
Cambodian households 200, 204
Cambodians 136, 196

Cambodian Socio-Economic Survey (CSES) 197
Cambodian villagers 136
capital 10, 26, 29, 30, 32, 33, 35–38, 113, 148, 162, 166, 183; accumulation 5, 25–27, 29–37; financial 155, 158, 160; mobilization 32–33
capitalization 10, 37, 38, 220, 225
caste-based occupation 25
castes 5, 7, 8, 36–38, 62, 64, 148, 164, 165, 170, 171, 218, 219; groups 3, 64
Castree, N. 45
Cattiyur, 62, 66, 67, 68, 94, 96, 99, 100, 105, 106, 108, 167, 168, 170, 172, 173, 175, 176
Chap, S. 182
Cha village 46–48, 52
Chhnok Tru Commune 132–144
Chhnok Tru village 132–144
children 10, 46–48, 50, 51, 85, 86, 92, 104, 108, 132, 133, 140, 141
Chinna Pattinavars 28
Church 108, 113, 122, 126, 149, 150, 153, 155, 156, 158, 160, 161, 171, 176, 178
citizenship 133, 134, 137, 221
civil war 58, 62, 72, 75, 76, 86, 88, 89, 92, 165, 222
claim fish resources 67
class 37–38; dynamics 25–38; relations, changing 30–35
climate change 1, 3–6, 43, 44, 58, 63, 67, 223
coastal areas 6, 11, 41–43, 52, 78, 155, 186, 188, 191, 196, 218
coastal communities 5, 6, 78, 192
coastal fishing communities 6, 13, 50–52
coastal land 189, 191–193, 220, 223, 225
coastal villages 50, 184
coast-to-coast migration 59, 92
Coleman, J.S. 166
collective action 3, 9–10, 13–14, 164, 165, 176–179, 183, 184, 219, 222–224
collective networks 9, 219
Colombo 173, 206
community-based organizations 98
community-based tourism 223
community fisheries (CFs) 14, 41, 42–43, 132, 181–185, 188, 190–193, 222–223; in Cambodia 181–183, 223; establishment of 14; gender and power struggle 181–194; Konkang Community Fishery 189–190; women 183–184; women leadership 190–192
compensation 124, 156, 158
conflict, fishers 58–73; conceptual approach 60–61
conflicts 2, 3, 11, 12, 27, 28, 37, 58–69, 71–72, 106, 182, 185, 191
consumption 26, 47, 118, 198, 206, 209–212, 215; data 198; distribution of 13, 18, 196, 198, 202, 204, 206, 208, 213, 215, 222; higher 210, 213, 216; monthly 198, 204; scale 204, 207, 209, 215
consumption expenditure: in fishing communities 196–216
contestation, fishers 58–73
cooperation 166
Cornwall, A. 191
Coulthard, S. 41, 94
crew members 28, 31, 33, 99, 118, 148, 153, 156, 157, 169, 170
Crona, B. 6
Cuddalore District 27–29, 33–34, 147–150, 153, 157–158, 161, 196, 197, 200, 205–207, 210–211, 213–216
cultural knowledge 166
cultures, traditional fishing 86, 87

data collection 16
debts 35, 38, 45, 47, 102, 103, 105, 106, 116, 149, 171, 173–174, 185; sources *160*
deep-sea fishing 13, 75, 88, 112, 113, 117, 118, 127, 197, 210, 213, 215
deep-sea shark fishing: in Thoothoor 113–116
Democratic Kampuchea period 41
density curve 211, 213, 214
density functions 209, 211, 212, 215; kernel 198, 209, 210, 213, *213*, *214*
depletion 2, 4, 11, 59, 60, 66, 67, 71, 96, 138, 140, 143
destination 27, 116, 149
distant waters, fishing 112–128; gendered identities and well-being 124–127; identity and mobility, labour 121–124; learning and developing skills 118–121; migration

and identity 116–118; risks and forbearance 117–118; risks bearing and learning skills 116–118; wealth and precarity 121–124
distress migration 155–157
Doss, C.R. 184
dowries 10, 33, 38, 82, 122, 150, 155, 158, 160, 161, 220, 223
dried-fish 97, 105; traders 170

earnings 32–34, 48, 51, 52, 81, 97, 105, 112, 125, 128, 198
east coast communities 14, 68, 164, 178
east coast fisheries societies 178
east coast fishing communities 176
ecotourism 189, 190, 192, 193
education 8, 45, 48, 52, 77, 122, 134–136, 150, 160, 166, 183
Ehrentraut, S. 134
empirical density functions 210
environmental changes 5, 6, 11, 26, 44, 79, 135, 162, 218, 223
Epstein, G. 167
ethnicity 37, 64, 77, 93, 94, 134, 148, 164, 165, 173, 176, 224
ethnic Vietnamese fishers: adaptation strategy 132–144; fishing households, coping strategies 136–143; immobility as disempowerment 134–135; immobility of 132–144; on Tonle Sap 133–135
ethnography 14
Exclusive Economic Zone 25
expenditure 32, 35, 125, 139, 198

Fabinyi, M. 61, 72
factory work 51
farming 34, 49, 50, 99, 104, 143, 168, 185; villages 34, 156
farming-fishing community: on floodplain 49–50
Felli, R. 45
female household heads (FHHs) 75–77, 79, 81, 88, 221, 222; fish vending, Batticaloa town 81–83; post-tsunami displacement 83–86; postwar period 86–87
female leadership: in K'dam Community Fishery 187–189
fish 42, 43, 46, 47, 49–51, 67, 69, 84, 87, 116, 120, 121, 139; collectors 47, 52, 136; fermented 136; flying 102; fresh 97, 186; labourers 12, 51, 85, 105; market 81, 82; movements 58, 65, 112, 165; prices 43, 46, 107, 171; selling 82, 187; species 7, 27, 93; stocks 60, 223
fisherfolk 3, 4, 8, 9, 11, 62, 219; in Global South 4; small-scale 222; *see also individual entries*
fisheries 5, 7, 10, 42, 44, 59, 60, 68, 69, 79, 94, 96, 176–178, 183, 187; assets 96, 97, 106; conflicts 68; governance 61, 166; inland 26; management 224; management policies 128; migrations 61; officers 69–71, 79; overfished 60; resource scarcity 96; *see also individual entries*
fisheries authorities 58, 69–71, 103, 104, 171; support of 69, 103
fisheries economy: in Puttalam District 59–60; in Trincomalee District 59–60
Fisheries Management Resource Centre (FishMarc) 197
fisheries resources 5, 6, 58, 66, 71, 72, 106, 182, 223, 224; declining 79; depleted 93; management and use of 181
fisheries sector 4, 5, 12, 42, 76, 193, 225; small-scale 1
fisheries societies 67, 69, 98, 104–105, 169, 176, 177; cooperative 169, 178
fishermen 28, 33, 66, 103, 124
fishers 2, 3–7, 10–13, 41, 44, 45–53, 63–67, 70–71, 112–119, 128, 161, 171, 174, 187, 223–226; commercial 43; deep-sea 123–125; east coast 68–70, 177; female 3, 9, 10, 16, 75, 84, 87, 218; full-time 42, 44, 46, 185, 186, 191; illegal 42, 182, 183, 186, 191; male 2, 33, 35, 108, 113–117, 119, 120, 122, 153, 154, 176, 177, 185, 220, 221; skilled 112, 221; sustaining 7, 93; traditional 25, 116, 157; west coast 63, 69, 102, 177
fishing 10, 11, 42, 46–53, 84, 87, 88, 122, 124, 138–143, 182, 186–188, 200, 205, 220; and agriculture 205, 209; camps 64, 170; communities in Tamil Nadu 197, 210; conflicts 42, 191; in distant waters 112–128; enterprises 36, 113, 125, 128, 153, 161; *see also individual entries*
fishing area 181, 189, 190, 192, 193; designated 14, 42

fishing boats 46, 63, 124, 169, 185; deep-sea 125; illegal 185; large 187; *see also* boats
fishing communities 1, 2, 5–7, 13, 41–42, 58, 64, 71, 94, 164, 196, 213, 219; in Cambodia 42–44; consumption expenditure in 196–216; livelihoods in 199–207; livelihood system and social relations 96–99; local 3, 218; migrant and host relations 58, 59, 62–65, 220, 222; small-scale 61, 75, 76, 109, 164; on Sri Lanka 92–109; on Tonle Sap lake 46–48; traditional 4
fishing equipment 26, 43, 48, 50, 120, 125, 185
fishing gear 63, 64, 68, 70–72, 105, 107, 148, 150, 153, 154, 156, 169, 171; destructive 36, 37; new 60, 169
fishing households 44, 46, 53, 60, 113, 164, 175, 176, 196, 197, 200, 204–216; consumption distribution of 206, 215; deep-sea 214; distribution of consumption 208–215; gender norms in 53; local 106; members of 41, 61, 93, 99, 167; migrating 99, 172; migration 208–215; mobility 208–215; multiday 153, 215; non-migrating 172; small-scale 218
fish resources 1, 2, 10, 11, 46, 47, 58, 60, 61, 65–67, 71, 140, 191; in Cambodia 41–54; changing 54; protecting 192, 222
fish traders 70, 103, 106, 171, 173, 178; sponsoring 107, 108
fish vendors 10, 79, 81, 82; female 81, 84
focus group discussions (FGDs) 114, 117, 167
formal networks 149
fresh-fish traders 170
Friend, R. 41, 44

gender 11–12, 37–38; gap 52; perspective 7; roles 2
gendered decision-making *126*
gendered household decision-making 92–109
gendered networks 164–179
gendered subjectivities 14
generation 37–38
Gioli, G. 44, 45
glocalization 225, 226
Guerin, I. 45

Hall, D. 78
Hapke, H.M. 7
Higgitt, R. 184
horizontal conflicts 71
host communities 63–65, 67
household decisions 107–108
Humphries, J. 76
Hun Sen 182

identity 12–13
income 32
in-depth interviews 27
India 3, 196–216; *see also individual entries*
intensive capital investment 27
International Centre for Ethnic Studies (ICES) 16

Johnson, D.S. 8
Joshi, B. 45

Kadalpalli 62, 65, 66, 68, 69, 73, 94, 96, 99, 100, 102, 103, 104, 105, 106, 107, 108, 167–169, 171–179
Kalpana, K. 2
Kanyakumari District 128, 147, 148, 150, 152–154, 156, 158, 197
Karaiyar caste 62, 64, 165, 171
Karava caste 64, 74, 165, 171
Kaufmann, V. 135
K'dam Community Fishery 184, 185–189; female leadership in 187–189; land grabbing, mangrove forests 185–187; seaweed cultivation company, conflict 187
kernel density functions 198, 209, 213, *213*, *214*
Khan, S.S. 184
Khmer people 132
Kraan, M.L. 93
Kurien, J. 181

labour mobility 27
labour mobilization 34–35
labour relations 35–36
Lawreniuk, S. 135
Liberation Tigers of Tamil Eelam (LTTE) 78

livelihoods: distribution of consumption 202–207; in fishing communities 199–207

Malthusian theory 93; of migration 96
Mamun, A. 183
Mann-Whitney test 150
marginal space 184
Marine Products Export Development Authority 25
market exposure 1
Marquette, C.M. 93, 109
McGregor, A. 109
mechanization 5
medium-scale fresh-fish traders 171
Meinzen-Dick, R. 184, 192
Meola, C.A. 183
migrant fishers 2; identity and well-being, issues 112–128
migration 2, 6–7, 10–11, 27, 88, 219; for capital accumulation 25–38; diversification strategy, fishers 44–46; industry 149; mobilizing for and against 164–179; social and gender relations in 148–150
mixed-method approach 222
mobile livelihoods 14
mobility 6–7, 10–11, 218–226
modernization 5
Morand, P.A. 44
Mortreux, C. 45
Mousset, E. 42
Mukkuvars 114, 116, 117, 126, 128, 148, 156

Nenadovic, M. 167
networks 13–14; of support 9–10
new economics of labour migration (NELM) 44, 45
Njock, J-C. 2

occupational profile, villagers 30
Overå, R. 7, 93, 109

Parsons, L. 135
Paravar caste 165, 171
Pattinavars 148, 157
People's Republic of Kampuchea (PRK) 134
Periyapattinam 29
Periya Pattinavars 28
Petchesky, R. 47
political ecology 61
Pollnac, R.B. 46
post-tsunami displacement 83–86; new settlement 84–86; old village, Ampara District 83–84
poverty 219, 224, 225
precarity 8, 12, 30–35, 77; mobility 147–162
privatization 5
Putnam, R.D. 166
Puttalam District vii, viii, ix, xviii, 14, 15, 19, 59, 60, 62, 64–67, 87, 93–98, 100–109, 112, 165, 167, 172–177, 197–200, 202–204, 206, 207, 212–216, 220

Radel, C. 44, 45
regionalism 63
relational aspects 9
resource access 58–73
resource degradation 223
Rigg, J. 8
risk-taking behaviours 113
Rosendo, S. 6
Rowlands, J. 184
Ruwanpura, K.N. 76

sawfish teeth 114
Schwartz, N.B. 43, 44
Scott, J.C. 184
seasonal coast-to-coast fisheries migration 58, 92
seasonal fishing 2
seasonal internal migration 58
seasonal migration 58–73, 99–107; benefits and costs 105–107; enabling and disabling factors 103–105; migrations patterns 99–102; reasons for 102–103
self-actualization 128
Selippur 62, 64, 65, 66, 67, 68, 69, 70, 71, 94, 96, 99, 100, 102, 103, 104, 108, 167, 168, 172, 175, 176
Sen, A. 77
shark hunters 112, 221
shark hunting 114
Sheridan, A. 184
Siar, S.V. 183, 192
Sinhala-Tamil Catholics 64
Siriwardena-de Zoysa, R. 64, 68, 166, 178
small-scale fisheries sector 1

small-scale fishers: in Asia 218–226; on Coromandel Coast, Tamil Nadu 25–38; kin solidarity and migrant networks 147–162; migration 93
small-scale fishing communities: in Eastern Province 78–79; social exclusion in 75, 77–78
Sneddon, C. 5
Sobrado, C. 198
social capital 148, 166, 167
social communication system 26
social exclusion: in small-scale fishing communities 75, 77–78
social identity 1, 6, 38, 88
social networks 149, 164, 166, 168–171, 178, 219, 221; for collective action 176–178; community organizations and institutions, relations 168–169; distress migration 155–157; family members, friends, and neighbours, relations 168; fishing and other employment, overseas 154–155; labour relations 169–170; livelihoods and migration, significance 171–176; mapping 16; market relations 170–171; mobilizing 176–178; multiday fishing 152–154; in seasonal migration 176–178; state and Church, relations 171; for women and men *175*
social polarization 2
social relations 148, 178
social well-being 61, 92–109
Sri Lanka 3, 5, 6, 196–216; conceptual approach 166–167; conflicting perceptions 65–68; fishing communities 92–109, 164–179; methods and study locations 167–168; Muslim fishers 69; regulatory environment and conflict 68–71; right to migrate *versus* local resources 65–68; sending and receiving districts, fisheries 94–96; social relations and networks 168–171; sociocultural and historical overview 165; Tamil fishers 69; west and east coasts of 58–73, 92–109
state policies 1
stresses 1, 2
subjective aspects 9
sustainable livelihood analysis (SLA) 44

Takizawa, S. 132
Tamil Nadu: gender relations 157–161; migration context in 150–152; small-scale fishers in 147–162; social landscape 157–161; well-being outcomes 157–161; west coast of 112–128
Teh, L.S.L. 44
Thmei village 52
Tiwari, P.C. 45
Trincomalee District vii, viii, ix, xviii, 14, 15, 19, 59, 60, 62–70, 73, 93–104, 107, 109, 165, 166, 167, 172, 175, 177, 197–200, 203–204, 206, 207, 212, 215, 220
tsunami 26, 28, 58, 63, 76

United Nations Research Institute for Social Development (UNRISD) 224
Uppanankuppam 29
Uppukadal 62, 66, 67, 69, 70, 71, 94, 100, 107, 167, 168, 172, 174–177

vulnerability 8, 12, 76, 77

wealth index 123, *123*
well-being 3, 8–9, 12–13; politics of 218–226
Wernersson, J. 45
Westlund, L. 2
White, S.C. 109, 167
women 66, 89, 97, 128, 135, 170; educated 128; empowerment 143; in fishing organizations 183–184; leadership 190–192; *see also* female household heads
women's livelihood 52
The WorldFish Center 46, 53

Xi, J. 46

For Product Safety Concerns and Information please contact our EU
representative GPSR@taylorandfrancis.com
Taylor & Francis Verlag GmbH, Kaufingerstraße 24, 80331 München, Germany

www.ingramcontent.com/pod-product-compliance
Ingram Content Group UK Ltd.
Pitfield, Milton Keynes, MK11 3LW, UK
UKHW021833260926
475702UK00049B/504

* 9 7 8 0 3 6 7 0 3 0 4 7 6 *